AF616203

THE MOVEMENT OF THE HEART AND BLOOD

Frontispiece. William Harvey

AN ANATOMICAL DISPUTATION CONCERNING THE MOVEMENT OF THE HEART AND BLOOD IN LIVING CREATURES

BY

WILLIAM HARVEY

TRANSLATED WITH
INTRODUCTION AND NOTES BY

GWENETH
WHITTERIDGE

MA, DPhil, FSA
HonFRCP

BLACKWELL SCIENTIFIC PUBLICATIONS
OXFORD LONDON EDINBURGH MELBOURNE

Osney Mead, Oxford
8 John Street, London WC1
9 Forrest Road, Edinburgh
P.O. Box 9, North Balwyn, Victoria, Australia

First published 1976

British Library Cataloguing in Publication Data
Harvey, William, b. 1578
An anatomical disputation concerning the movement of the heart and blood in living creatures
Bibl.—Index
ISBN 0 632 00059 7
1. Title. 2. Whitteridge, Gweneth
612'.1 QP102
Blood Circulation—Early works to 1800

Distributed in the United States of America by
J. B. Lippincott Company, Philadelphia
and in Canada by
J. B. Lippincott Company of
Canada Ltd, Toronto

CONTENTS

LIST OF ILLUSTRATIONS

PLATES

FIGURES IN THE TEXT

PREFACE

On translating De motu cordis

It would be both foolish and erroneous to pretend that this is a totally new translation of *De motu cordis*. It was translated into English for the first time by an unknown writer and published in London in 1653. It was reprinted once in 1673 and reissued, edited by Sir Geoffrey Keynes, in 1928 to mark the tercentenary of the first appearance of *De motu cordis*.* In 1847 it was retranslated by Thomas Willis, a learned country doctor, and published for the Sydenham Society in *The Works of William Harvey*. For more than a hundred years this remained as the standard English version and in 1907 it was reissued as a volume in the Everyman's Library series. In 1957, to mark the tercentenary of Harvey's death, Dr K. J. Franklin made a new translation into modern English and this was published from Oxford as a volume in Blackwell Scientific Publications, and in 1962 was adopted to replace Willis's version by Everyman's Library. In addition, in 1928, Dr Chauncey Leake published in America a new rendering of the text based on Willis's translation but this has never been easily available in this country. It is, however, the only edition to contain any explanatory notes to the text or to have an index.

The first translation is written in the language of Harvey's own time. Its vocabulary and turn of phrase is what one might have expected of Harvey himself had he written in English.

* A 'modernized' version of the 1653 edition was published by Dr Michael Ryan in the *London Medical and Surgical Journal*, volumes 1 and 2 (1832–3). The text is scattered through the volumes as if to fill up gaps between the articles. The version is of no particular merit.

But because of this its meaning is now not always immediately clear and it does contain a few actual mistakes and a certain number of omissions. It has not till now been realized that it is not based on the 1628 text, but on that of the revised edition of *De motu cordis* printed in 1648, in which the editor has tried to correct a number of the misprints of 1628. Willis's translation of the 1628 text is written in a style that is characteristic of the nineteenth century. It is verbose where Harvey is brief, heavy and flat where Harvey chooses a colourful and vivid phraseology so that it misses a great deal of the vigour of the original. Chauncey Leake succeeded in removing some of the heaviness, but at the expense at times of wandering rather far from the Latin in what was a 'deliberate attempt to present Harvey's thought in the current physiological manner'. Franklin's version went even further in this direction with the result that it is at times so free that it is no longer a translation but a paraphrase. Moreover, the translator did not have a sensitive ear and the language which he uses is totally out of keeping with Harvey's style both in his English and in his Latin.

Having carefully considered all these versions, I came to the conclusion that the first translation of 1653 was undoubtedly the best, in spite of its various shortcomings, for it alone had any real feeling for the style of the Latin on which it was based. The introduction of modern semi-scientific phraseology of the kind which Franklin employs does not represent Harvey's thought for it is anachronistic and sometimes implies a precision which could not possibly have existed in the seventeenth century. Willis's long-windedness, though usually accurate, is quite unnecessary and again hides the quality of the original. I therefore decided that I would model my translation deliberately on the first English version using the reprinted edition of 1928 and checking it against both the Latin texts. In this way I have tried to keep the flavour of the seventeenth-century English where it reproduces with marvellous clarity the vigour of Harvey's Latin phraseology. But at the same time I have tried to avoid using any word whose meaning is not immediately obvious and any turn of phrase that might be considered archaic. I find a directness and simplicity in Harvey's writing which I have done my best to convey in the English. I have tried to translate and never to paraphrase. I have looked carefully at Harvey's deliberate choice of words where synonyms were available to him and have chosen as well as I could the English equivalents best suited to the context. I have not made use of any modern technical phraseology in matters anatomical or physiological. The only exception to this is the use I have consistently made of pulmonary artery and pulmonary vein for the seventeenth-century arterial vein and venous artery. (In one passage quoted by Harvey from Galen I have kept the old terminology.) It requires a perpetual effort of thought on the part of anyone today to remember that before the general acceptance of the circulation of the blood the pulmonary artery was considered to be a vein and the pulmonary vein an artery. There are only one or two places in the text, where Harvey is explaining the differences between them, that this use of modern terminology results in a somewhat peculiar reading of the English, and for these instances I make no apology.

One or two other words call for comment. One is 'nervus' which in the seventeenth century could still be used for nerve, sinew or fibrous tissue and when it appears in the Latin I have tried to give the meaning appropriate to the context, but in some cases there is a little doubt as to what that should be. This also applies to the adjectives derived from it. With regard to the lists which occur from time to time of animals that Harvey had used, I have not found it possible to be certain what he means by *conchae*, whether *astacus* is really a 'crayfish' and whether *squilla* is always a 'shrimp', or if not what he meant by 'squill-fish'. He was writing before the introduction of any standard classification into zoology and the explanation of these words provided by the dictionaries of Classical Latin is not very helpful.

Finally, in addition to the English translations to which I am indebted even when I have rejected their readings, I am also indebted to the translation made into Italian by Dr Loris Premuda (Milan 1957), for I found that his translation of the chapter headings and some of the difficult passages were more accurate than those given in any of the English versions.

ACKNOWLEDGEMENTS

I have acquired debts of gratitude to various people, among them, to my husband, Professor David Whitteridge FRS, for his patience in discussing Harvey's physiology and anatomy, to Mr Colin Hardie, late Public Orator in the University of Oxford, for his advice on Harvey's use of Latin, to Mr Douglas Fisher FRPS, for providing me with the photographs from which the plates in this book are made, to Professor Geoffrey Dawes FRS, for giving me his diagram of the foetal circulation in the sheep. All mistakes in Latin, physiology or anatomy are my own. I should like to thank also my publishers for their kindness and help in technical problems and for obtaining permission from the Oxford University Press to reproduce figures 2, 3, and 4, published by them.

INTRODUCTION

1. THE STRUCTURE OF *DE MOTU CORDIS*

I have proceeded according to the strictest form

Although it is common nowadays to refer to *De motu cordis* in this abbreviated manner, it is of some importance to bear in mind that its proper and complete title is *Exercitatio anatomica de motu cordis et sanguinis in animalibus*, for this immediately explains its structure. It is an *exercitatio anatomica*, an 'anatomical exercitation'. According to the definition given in the *Shorter Oxford English Dictionary*, an exercitation is a 'display of skill' in written or spoken argument in support of a particular thesis. In other words, it can stand for a 'disputation', such as is still required in many European universities of those who 'dispute' a thesis in order to obtain a Doctor's degree. At the present time it is customary for the first announcement of any discovery in the biological world to be made in the form of an article in a scientific journal and, moreover, for this article to be written in accordance with a set pattern and form which has been established as being that in which the validity of the experimental evidence on which the new hypothesis is based can be most easily appreciated and assessed, and consequently compel most readily the agreement of contemporaries. As no such periodicals existed in Harvey's time, it would, therefore, seem most likely that he would adopt for the purpose of announcing his discovery of the circulation of the blood, that form of literary exercise which was recognized by his contemporaries to serve the same purpose as the modern scientific article. And this form was that of the disputation.

When Harvey was a student in Padua, disputations were thought to be at least as important as lectures, if not more so, for the training of the future physician and all medical students were expected to attend them regularly. Every professor who taught medicine or philosophy as part of the medical curriculum was required to hold a public disputation in the Schools at least twice a year, once before and once after Easter. If he failed to do so he was liable to incur a fine of 50 lire. As there were then about twenty professors in the various subdivisions of these two subjects, it would seem that there must have been at least one such formal disputation each week during term. While these disputations were taking place, no formal lectures were given. They were occasions for a certain amount of ceremonial and they were conducted on strictly formal lines governed by a series of rules. The subject for debate was chosen by the professor and the proposition was then formulated and upheld by a series of arguments advanced in strict logical form, that is in accordance with the various figures of the syllogism, a fact which explains why all students had to attend lectures in formal logic in their undergraduate years. As many as seven students could take part in each disputation, but any student wishing to dispute had to give prior notice of his intention to do so, and had to be of at least two and a half years' standing. To have taken part in such a disputation was a necessary prerequisite for the obtaining of a Doctor's degree for which part of the final examination consisted in an exercise of this kind before a jury of the teaching members of the Faculty. The students' part in the practice disputations consisted in producing refinements in the argument, making logical distinctions and in this manner supporting or countering the premisses as they were produced in the course of the debate. When the conclusion had been reached, the verdict was given by the professor. In addition to these formal disputations in the Schools, informal ones also took place in the evenings, after the close of the academic day. They were known as the *disputationes circulares* and were held in various places within the city of Padua. Every professor of medicine or philosophy was required to attend these evening disputations for at least half-an-hour each day and there to discuss with the students the statements he had made in his lectures, listen to their objections and resolve their doubts. Any student wishing to propose a new solution to any problem under dispute might do so on these occasions, provided that he had previously obtained permission. These less formal disputations were presided over by the Rector of the University of the Artists, who was himself a student, and in them the debate was also conducted in accordance with the established rules of argument.

Formal and informal disputations of this kind formed part of the curriculum for medical students in every European university in the sixteenth and seventeenth centuries.* The Laudian Statutes of Oxford, promulgated between 1630 and 1636, commended the practice and it was only during the course of the eighteenth century that it fell into disrepute when the debate became more formal than real, sometimes consisting in nothing but a series of set questions and set answers, the subjects discussed being little more than trivialities. For Harvey's contemporaries, however, the medical disputation was a serious attempt to explain and account for diverse phenomena in accordance with the accepted rules of argument.

* A manuscript in the Bodleian Library, Rawlinson D 213, contains summaries of disputations held in Cambridge between 1592 and 1597, mostly on medical topics. One is on the thesis that the heart is not the source either of the veins or of the nerves. (I am grateful to Dr Vaisey for calling this manuscript to my attention.)

Harvey's choice of the medical disputation, therefore, as the form in which to set forth to the medical and scientific world in general his discovery of the circulation, was in accordance with the practice of the time. When, in his *Anatomical Lectures*, he is about to announce his conclusion that the timing of the systole and diastole of the heart is the reverse of what is commonly supposed he says: 'This error has been current for a very long time and therefore I have proceeded according to the strictest form because the subject is so ancient a one and has been studied by so many great men' (p. 267). There follows a series of observations and arguments which can be based upon them and the statement: 'This can be proved by actual inspection in the living and in the dead, by reason and by experiment' (p. 269). His 'strictest form', therefore, is reasoned argument based on observation and tested by experiment; but here there is no question of any formalizing of the manner of the argument. This appears in the Epistle Dedicatory to the College of Physicians in *De motu cordis*, where he says in words similar to those of the *Lectures*:

> And since this book alone affirms that the blood goes forth along a new path and returns again, contrary to the way which has been accepted for so many hundreds of years by countless most famous and learned men and by them followed and well worn, I was greatly afraid to suffer this little book . . . to come forth into the public view . . . unless I had first laid it before you, proved it by ocular demonstration, answered your doubts and objections and received the verdict of your most accomplished President in my favour.

From the end of this sentence, therefore, it is clear that Harvey had 'disputed' his thesis with the College of Physicians and at its conclusion the President had pronounced the verdict. *De motu cordis*, therefore, is the writing up of this disputation with the difference that in it Harvey himself raises the objections and answers them as the debate proceeds.

That the form of the disputation underlies *De motu cordis* is most easily to be seen from the headings of Chapters 9–17. Having announced in Chapter 8 the general hypothesis that there is a circulation of the blood, in Chapter 9, following the usual procedure of a disputation, Harvey resolves, or divides, his main thesis into three so-called 'hypo-theses', or propositions.

> But lest anyone should say that I am speaking empty words and making fair assertions only without foundation and introducing a new thing without just cause, there are three propositions which come now to be proved. . . .

These three propositions are as follows.

1. That the pulse of the heart transfers the blood incessantly from the vena cava into the arteries in so great a quantity that it cannot be provided by the food eaten and in such a way that the whole quantity passes through the heart in a short time.
2. That the pulse of the arteries drives the blood into every part of the body in a quantity that exceeds the amount contained in the whole body and in far greater quantity than is necessary for nutrition.
3. That the veins uninterruptedly bring back the blood from every part to the heart.

If for the moment we defer the question as to how these propositions are treated in the following chapters and continue to look at the chapter headings, it will be seen that two chapters are devoted to each of the first two propositions and one to the third, and that the section ends with a formal statement of the conclusion. Chapters 9–14, therefore, constitute the demonstrative proof of the circulation of the blood organized in a formal manner from the announcement of the hypothesis to the statement of the conclusion. A demonstrative proof is one that is based on observation of phenomena tested by experiment, and its validity as a method of scientific enquiry and formal argument had been recognized since Galen's time.

This demonstrative proof is followed in Chapter 15 by a different kind of proof, namely that derived from logical argument alone.

> It would not be irrelevant to the matter in hand to add these arguments also, because, according to certain common processes of formal reasoning, a thing is so if it be both accordant with reason and necessary.

These two terms, *conveniens* and *necessarium*, are used by Harvey in the sense in which they had appeared in arguments conducted according to the set rules of logic at any time from the days of St Thomas Aquinas. That which is *conveniens* is in accordance with reason. To demonstrate an argument logically is to show a uniformity in the sequence of cause and effect which is involved in the very nature of things. Harvey begins by asking acceptance for the Aristotelian maxim that dead things are cold and living things are warm, that there must be a centre of warmth in living things and that this centre is the heart. From this he argues that the blood leaving the heart conveys warmth to the extremities where it becomes chilled and therefore must return again to the heart to reacquire warmth and so preserve life in the whole body. For proof he refers to the observation of hands that are blue with cold, and warmed again by the influx of blood which losing its own heat in them must return again to the source of heat. When the heart is impaired for any reason, the outflowing blood cannot repair the deficiency of heat. Turning next to the Aristotelian corollary that the heart is the source of perfected aliment, Harvey starts from the premiss that it is the only part of the body to have blood both for public and private use (that is, for itself and for the use of the rest of the body), and therefore it must be able to distribute the blood; therefore it pulsates and drives the blood for fair distribution among the parts. Blood, like spilt water, tends to coalesce and return to the greater mass, therefore blood from the periphery will tend to return to the centre from where the pulse forcibly drives it out. The events in both series are alike necessary and conformable with reason.

In Chapter 16 Harvey turns to yet another kind of proof, the proof *ex consequentiis.*

> There are various questions arising from this truth which has been postulated as consequential upon it, and taken as arguments *a posteriori* are of some value in confirming its validity.

Harvey now uses the form of argument known to the Greeks as an enthymeme, that is a syllogism in which the middle term is not expressed. Instead of starting from a statement of fact, this kind of syllogism, also known as an hypothetical or conditional syllogism, can start from an hypothesis, in which case the truth of the conclusion is conditional on the truth of the hypothetical first premiss. It is included among the forms of syllogistic argument described by Niccolò Massa in his *Logica*, printed in Venice in 1550, and specially designed for the use of 'university students of philosophy, medicine, rhetoric, grammar and history'. Harvey is here dealing with a series of phenomena which in his own time appeared to be 'enveloped in much ambiguity and obscurity', but for which the explanation might lie in his hypothesis of the circulation. In this chapter, therefore, we find a different approach to the problem. In the first part of the book and in setting out the demonstrative proof, Harvey has tested the various subdivisions of his hypothesis to see whether each in its turn accounts for certain observed phenomena, that is to say, he has proceeded according to what we call the hypothetico-deductive method. And in each case he has shown that the consequences which can be deduced from the hypothesis do indeed correspond to observed reality. In this Chapter 16, he is using the hypothesis to predict that which is not yet known. Whereas today this is recognized as an important process in the validation of an hypothesis considered as a temporary stage in the course of an investigation, in that it opens up the field of enquiry and points to the possible designing of further experiments by which the hypothesis may be tested and so increase its credibility or require its modification, in Harvey's time it was a method which was not much used as part of a biological investigation. Its appearance, therefore, in *De motu cordis* is of considerable interest and importance. From it derives Harvey's vision of the many and diverse phenomena in the whole field of medicine that were inexplicable to him and to his contemporaries but which might have an explanation in his hypothesis of the circulation. In the light of our present knowledge, not all the phenomena which he describes are to be accounted for as he thinks, but his prediction provides a temporary and pragmatic explanation, the best possible one for his time.

In Chapter 17 Harvey continues with the same form of argument based on the conditional syllogism. He turns away from prediction as such to a more limited exposition of proof *a posteriori*. He discusses only those phenomena which can be observed during dissection and vivisection and whose explanation seems to lie in his hypothesis. His reason for so doing he had already given in the preceding chapter:

> it is my wish that it be established and supplied above all else with arguments based on anatomy.

His observations, therefore, are used in this chapter as a colligation of facts, that is of facts apparently disconnected but linked together by an hypothesis which reasonably accounts for them all.

All these phenomena which can be observed during dis-

> section and a great many more, if they be rightly weighed, are seen to shed light abundantly on the truth that I have stated and to prove it completely . . . for it were very hard for anyone to explain by any other way than I have done for what cause all these things were so made and appointed.

If we now turn to the first part of the book, leaving aside the first chapter which merely explains why Harvey wrote it and has nothing to do with the structure of the whole, it will immediately be seen from the chapter headings that Chapters 2–7 consist of the exhaustive and systematic description of observed anatomical facts and physiological phenomena which will form the basis for the formulation of the hypothesis in Chapter 8, the evidence on which it will rest. Because the correct timing of the systole and diastole of the heart was not then generally recognized, it is from the analysis of the movements of the heart that Harvey begins. This leads him logically to the examination of the movement of the arteries, the timing of whose systole and diastole depended on that of the heart and was equally misunderstood by his contemporaries. He then turns to the problem of how the blood passes through the heart, the movements of the auricles and then of the ventricles, and so comes in Chapter 5 to the distribution of the blood through the aorta to the whole body. Throughout, his statements are based on observation and tested by experiment. Each chapter is as it were a disputation in miniature. This is particularly clear in Chapter 2 where he begins with a series of observations on which he bases the proposition that when the heart contracts its walls are thickened, its ventricles narrowed and the blood is thrust out, a proposition which he says is conformable to reason, *rationi consentaneum est*. He proves the proposition by an observation and an experiment and passes on to the statement of a formal conclusion concerning the timing of the systole and diastole of the heart. He then raises an objection to his conclusion by introducing Vesalius's analogy of the bunch of rushes which he proceeds to refute from his knowledge of the anatomy of the fibres of the heart and his own analogy with the contraction of muscle. (Only in Chapter 17 does he reveal how he obtained this anatomical knowledge.) In the same way Chapter 3 begins with an observation on which a proposition is founded and it is supported by experiment and by further observation.

In Chapters 6 and 7, Harvey is concerned to prove the pulmonary transit of the blood which he does in Chapter 6 not by any anatomical demonstration but by arguments founded on observations in comparative anatomy and the logical inference to be derived from them; in Chapter 7 by quoting the passages from Galen's works in which the competence of the valves of the heart is stated and supported by argument. This chapter can be seen as having its parallel in Chapter 15 and perhaps Harvey chose to use this method here for the same reason, to convince those who would not believe anything unless an authority were alleged for it. There is no doubt that all anatomists of Harvey's time knew well enough these passages from Galen, but the whole theory of the breeding of vital spirit in the left ventricle of the heart, of the entry of air into it through the pulmonary vein and of the return through this same pulmonary vein of fuliginous vapours and of 'vital' blood made complete nonsense of Galen's statements, for the theory imposed the belief that the valves did indeed let something leak back past them. Harvey's own certainty of the competence of the valves of the heart rests, as is to be expected, on his detailed observation of their action, but his description of this he leaves until Chapter 17.

The structure of *De motu cordis*, therefore, is very simple. Chapters 2–7 set out in detail the anatomical evidence on which the hypothesis will rest; Chapter 8 summarizes the evidence and states the hypothesis which seems to account for the evidence; Chapters 9–17 prove it in three different ways: 9–14 by demonstration, 15 by logical argument, 16–17 by arguments *e consequentia* and *a posteriori*. The Introductory Discourse, though outside the main structure of the book, shows the same rigorous argument based on syllogistic form, as is apparent in Harvey's use of the accepted terminology of the disputation: *probabile*, *improbabile*, *conveniens*, etc.

In a few of the arguments which Harvey uses in this book there is no more validity than in many such arguments produced by his contemporaries. These arguments are not flawed by reason of any errors committed in the logic, but because the premisses which form the bases from which they start are themselves merely incapable of proof or erroneous. A case in point is his discussion of the passage of chyle through the mesenteric veins in Chapter 17. Here Harvey has made no direct observations on the differences between the venous blood in the veins from the stomach and that in the veins from the large intestine but he asserts that these differences exist. But Harvey himself realized that the validity of his hypothesis of the circulation did not rest on arguments of this kind. His predictions of Chapter 16 do not of themselves compel acceptance of the hypothesis, and it was because he understood that it was not profitable to use the unknown to support the incompletely known that he stresses the fact at the end of this chapter that the proof of what he has said rests on the anatomy of the body, animal and human; that is, on arguments that are verifiable and which anyone who choses can verify for himself.

Returning then to his demonstrative proof as set out in Chapter 9–14, it will be seen that he devotes two chapters each to the first two propositions, but that the effective proof of both of these is the argument from quantity which he advances in

Chapter 9 and to which he reverts for refutation of the objections which he raises in the course of the discussion and for the proof of his thesis. Chapter 10 clearly proceeds in the manner of a disputation by refutation of objections. Chapters 11 and 12 repeat again the quantitative argument, but his experiments with ligatures demonstrate conclusively the flow of blood outwards from the heart through the arteries. Chapter 13 is entirely devoted to his third proposition that the blood returns through the veins, and it consists in the demonstration of the action of the venous valves and of the direction of the blood flow in the veins.

*

It was suggested in 1928 by Dr Chauncey Leake that *De motu cordis* was written at different dates, an opinion which he had formed from differences which he thought he detected in Harvey's style in different parts of the book. He did not, however, pursue the question. Recently Jerome Bylebyl (*Bull. Hist. Med.*, XLVII, 1973, 427–70) has advanced the theory that *De motu cordis* is based on an earlier treatise written before Harvey knew the circulation and subsequently adapted for publication in 1628, but he has failed to find any conclusive evidence to support this opinion.

There is no doubt that much of the evidence which Harvey uses in *De motu cordis* he had accumulated over a great many years. The text of his *Anatomical Lectures* of 1616 proves this. But this text also proves that he had not then formulated the hypothesis of the circulation and that his statement in *De motu cordis*, Chapter 8, to the effect that everything that he has said so far was in fact known to Galen and Columbus, was true. Before he discovered the circulation Harvey had nothing beyond a greater clarity of thought to add to the knowledge of his predecessors. Only when the hypothesis dawned and the proofs came to hand had he any new contribution to make to the subject.

If we now turn to the *Anatomical Lectures*, it is quite clear that there are certain differences in the style of writing in different passages in the notes. The explanation lies in Harvey's constant habit of copying out verbatim passages of longer or shorter length from the various books which he was reading. These passages are embedded in his notes among his own remarks, observations and longer passages of discussion and debate. For the actual writing of *De motu cordis* it is reasonable to suppose that Harvey consulted the notebooks which he had accumulated over the years and in which he had recorded his observations, his experiments and his reading. From these he selected the relevant material but in so doing he collected together not only his own original comments and discussions but sometimes those of other people which he had also written down. Some of these have become embedded in the text of *De motu cordis* itself. I have pointed out in the notes one or two such passages, but I have made no deliberate search for more.

It is, furthermore, characteristic of Harvey that he has difficulty in keeping to the point. The whole of the text of *De generatione* show this abundantly. His knowledge is so wide-ranging over so many different fields that he is constantly bringing in some fascinating point suggested to him by something he has just written. This he does from time to time in *De motu cordis*, but the digressions are fewer in number and more nearly on the subject in hand than they are in *De generatione*. In this book some of his finest passages are his digressions and they are quite different from the style of much that he wrote in it elsewhere. But in both books the main argument is pursued with the same relentless application to every detail, even at the cost of many repetitions of the same points but in slightly different contexts. The underlying structure of *De motu cordis*, however, is more easily recognizable than that of *De generatione*. In the one Harvey's observations had led to the formulation and proof of an hypothesis which he could then set in the appropriate form before his colleagues. From the very nature of the problem he had set himself to solve, he could do no such thing with all the multitude of his observations concerning the generation of living animals.

De motu cordis, then, is the finished product, the writing up of work which had covered a great many years. It is not an account of how the work was undertaken, in what order the observations were made, nor does it reveal at what stage the one hypothesis which could explain them all first came into Harvey's mind. That it is, nevertheless, possible to attempt such an enquiry, using the evidence of the *Anatomical Lectures* and of *De motu cordis*, I will try to show in what follows.

2. HARVEY'S DISCOVERY OF THE CIRCULATION

Because I do not think it possible to reach the truth from other men's opinions whether they be given out on bare authority or even confirmed by arguments from probability, without the help of diligent personal experience, I will expound from the book of Nature . . .

The only independent testimony which we have concerning the way in which Harvey came to the discovery of the circulation is provided by Robert Boyle. In two different places in his books, Boyle says that Harvey told him 'in the only discourse I had with him', that it was the consideration of the valves in the veins that first 'hinted' to him, or 'induced him to think of', the circulation of the blood throughout the whole body.

> When he took notice that the valves in the veins of so many several parts of the body, were so placed that they gave free passage of the blood towards the heart, but opposed the passage of the venal blood the contrary way: he was invited to imagine that so provident a cause as Nature had not so placed so many valves without design: and no design seemed more probable than that, since the blood could not well because of the interposing valves be sent by the veins to the limbs, it should be sent through the arteries and return through the veins whose valves did not oppose its course that way.
>
> (*A Disquisition about the final causes of things*, London 1688)

I can think of no valid reason for rejecting as untrue Boyle's account of this conversation with Harvey. As he refers to it twice, it is to be presumed that it made a considerable impression on him and that he himself believed in the truth of what Harvey had said. Consequently, it is appropriate to examine

Harvey's writings closely, paying particular attention to this question of valves, to see whether the writings do or do not bear out the truth of this statement.

Fabricius says that he discovered the valves in the veins in 1574. In the winter anatomy of 1578–9, Caspar Bauhin, who was then a medical student at Padua, says that he saw Fabricius demonstrate them and when he published his anatomical textbook in 1590, Bauhin remarks in it that Fabricius had then continued to demonstrate the venous valves for sixteen years. Nine years later, in the autumn of 1599, when Harvey came to Padua, it is very likely that he too saw Fabricius's demonstration and heard his explanation. In 1603, just after Harvey had left Padua, Fabricius published his account of the venous valves and his explanation of their use:

> My theory is that Nature has formed them to delay the blood to some extent, and to prevent the whole mass of it from flooding into the feet, or hands and fingers, and collecting there. Two evils are thus avoided, namely, undernutrition of the upper parts of the limbs, and a permanently swollen condition of the hands and feet. Valves were made therefore, to ensure a really fair distribution of the blood for the nutrition of the various parts.
>
> . . . valves are placed in the veins less with a view to causing a pooling and stopping of blood before the oblique mouths of the branches, than with a view to checking it on its course and preventing the whole mass of it from slipping headlong down and escaping. . . . Nature has so placed the valves that in every case the higher valves are on the opposite side of the vein to those immediately below them. . . . In this way the lower valves always delay whatever slips past the upper ones, but meanwhile the passage of the blood is not blocked.

As is so often the case with Fabricius, his anatomical description

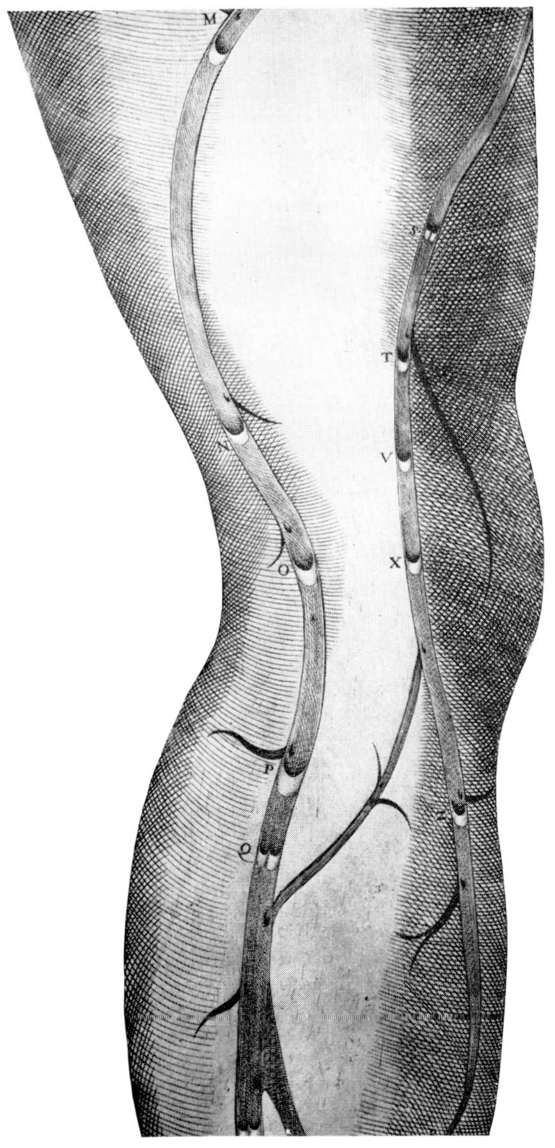

Plate 1. The valves in the veins as drawn by Fabricius

of the valves is reasonably accurate, but his interpretation, which is in accordance with the Galenic theory of the movement of the blood, is erroneous.

In the *Anatomical Lectures* of 1616 there is no description of the venous valves. Nor is there any suggestion that Harvey intended to mention them even in his account of the veins of the leg (which comes in his lecture notes for 1619), although most of the contemporary anatomical textbooks include them at this point. In all these notes Harvey only mentions the valves once, and he then ascribes to them a use which accords with precirculatory thinking, 'the valves set in a contrary direction break off the pulse both in the heart and in the other veins. WH and for this reason the veins have very many valves opposed to the heart' (p. 273). Clearly, he had not then begun to think that they might have some different use or to realize their importance.

In these *Anatomical Lectures*, however, we do find Harvey's first discussion of Columbus's work on the timing of the systole and diastole of the heart and his conclusion that it is the reverse of what was generally supposed. This work Harvey had repeated in great detail on a large number of different kinds of animals. 'I have observed these things for hours together and I have not been able easily to distinguish between them either by sight or by touch' (p. 265). But having analysed the heart's movement in detail, he comes to the conclusion that Columbus was right. But whereas Columbus's conclusion, as far as we know, was based entirely on observation, Harvey had performed the experiment of making a deep cut in the ventricle and watching the blood spurt out at each contraction. In addition, Harvey had by this time satisfied himself as to the truth of Columbus's views concerning the pulmonary transit of the blood. Because, however, it has not been sufficiently clearly understood that Columbus's hypothesis turns on the competence of the valves of the heart and not on any of his ancillary observations, it has not been noticed either that Harvey was in no doubt as to what was its essential feature. Having discussed the whole business at some length, Harvey sums up his conclusions in the *Lectures* in these words:

> From these things it is evident WH that the action of the heart in so far as it is an instrument of movement is to send the blood from the vena cava into the lungs through the pulmonary artery and from the lungs through the pulmonary vein into the aorta (p. 271).

The things from which it is evident are: the valves which prevent the return of blood to the heart, the pulse of the aorta and pulmonary artery but not of the pulmonary vein (also observed by Columbus) and the simultaneity of the systole of the heart and the arterial pulse. Clearly, therefore, before 1616, Harvey had investigated for himself the action of the valves of the heart and had come to the conclusion that they were competent. But he does not in the Lectures provide any first-hand description of these valves. He merely quotes briefly the words of Bauhin's textbook. For his own original observations we have to turn to *De motu cordis*, Chapter 17.

This chapter, whose purpose Harvey explains is to show how the anatomical evidence is the basis on which his hypothesis rests, contains a number of observations which Harvey had already made by 1616, as we know from the text of the *Lectures*, but of whose ultimate significance he was not then fully aware. It is in this chapter that we find Harvey's description of the valves of the heart.

> Consider the use of the sigmoid valves which are so made as to prevent the blood once sent forth from returning again into the ventricles of the heart. They are situated in the orifice of the pulmonary artery and in the aorta, and when they [that is, their cusps] are raised up, they meet

> together in a three-cornered line such as is left by the bite of a leech and so when they are tightly pressed together they prevent the return of blood.

The analogy with the bite of a leech is a totally original observation. It is a visual analogy and could only have been made by Harvey after he had actually seen the valves in action. He continues with a description of the mitral valve at the base of the pulmonary vein:

> In the left ventricle, so that there may be a more exact closure to withstand the greater driving force, there are only two valves [that is, cusps] shaped like a mitre and so that they may shut most precisely, they stretch out for a long way through the midst of the ventricle into its apex. . . . It is likewise certain that it was to prevent the blood from slipping backwards into the pulmonary vein and thereby impairing the strength of the left ventricle in driving on the blood into the whole body, that these mitral valves much surpass in size and strength and precision of closing those which are placed in the right ventricle.

Observations of this kind, if not these very ones, Harvey had already made by 1616, for he was then of the opinion that the papillary muscles and the chordae tendineae strengthen the movement of contraction in the heart (*Lectures*, p. 259). What is certain from the evidence of the *Lectures* is that by 1616 Harvey had repeated the whole of Columbus's work and verified his findings. It would, therefore, seem to follow that Harvey's own work on the movement of the heart and blood, which culminated in his discovery of the circulation, began from the verification of Columbus's discoveries.

That the whole theory of the pulmonary transit of the blood rests on the question of the action of the valves of the heart and not on the solidity or otherwise of the interventricular septum, is clear from what Harvey himself said in the letter he wrote to Paul Marquand Slegel in 1651, after he had described the experiment by which he proved conclusively that the septum was a solid wall, an experiment which he had only then lately devised, that is some twenty or more years after the publication of *De motu cordis*: 'I have not anywhere set this up as a fundamental basis of my circulation of the blood.'

In *De motu cordis*, Harvey does not waste time in re-proving the pulmonary transit of the blood. When in Chapter 7 he comes to use it as part of the evidence pointing towards the general circulation of the blood, he becomes Columbus's advocate and invites acceptance of the theory on the analogy of the percolation of the blood through the parenchyma of the liver, in which he says all men believe.

> Why then should they not likewise believe this of the passage of the blood through the lungs in men come to age, upon the same arguments? And with Columbus, a most skilful and learned anatomist, believe and assert the same. . . .

And then, saying that there are some who will believe nothing except when it is supported by authority he quotes the long passages from Galen to show that Galen had really taught that the valves were competent. It would be clean contrary to everything that we know of Harvey to suppose for one moment that he himself was convinced by this argument. His own certainty rested on detailed anatomical examination of the structure of the valves of the heart and observation of their manner of action, just as he describes in Chapter 17. And this was supplemented by the additional experimental evidence provided by his work on animals with no lungs and a single-chambered heart, and on the foetal heart, all of which he describes in Chapter 6. Some only of this work is referred to in the *Lectures* and by 1616 it had probably not been completed.

It seems likely, therefore, that Harvey's initial acceptance of Columbus's theory rested on his investigation of the valves of the heart.

Now bearing in mind Harvey's remarks to Boyle concerning the valves of the veins, it would not seem unreasonable to suppose that it was his work on the valves of the heart that sent him back to examine afresh the valves of the veins. His findings concerning them and his experimental demonstrations of their manner of action are all set out in Chapter 13. In this chapter there are many similarities with the treatise that Fabricius had written concerning the venous valves, and certain significant differences. It would seem, however, that as with Columbus, so here again Harvey started his own work with Fabricius's very much in mind. It would also seem reasonable to suppose that having satisfied himself as to the competence of the valves of the heart, the first question which he might ask of the venous valves concerned their competence. If the valves of the heart act in such a way as to allow the movement of blood through them in one direction only, do the valves of the veins act in a similar manner? A partial answer to his guess that this is so is furnished by a purely anatomical demonstration:

> I have very often found in dissecting veins that if, beginning from the root of the vein, I put a probe towards the small branches, with all the skill I could, it could not be further driven by reason of the hindrance of the valves. But on the contrary, if I put it from the outer branches towards the root, it passed very easily.... [The valves] yield a passage to a probe inserted from the outside inwards, and like flood-gates which stay the course of rivers, are most easily pushed aside.

Examining further the precise siting of the valves in the veins, Harvey makes two further observations. The first contradicts Fabricius's statement that their arrangement is such that they do not totally occlude the lumen of the vein. They are, Harvey says, mostly situated in pairs, looking towards the heart and looking towards one another and touching one another so that their tips are apt to stick together and be joined, and in this way they completely prevent any blood from passing through. The second observation counters the notion that the function of all the valves is to prevent the blood from falling down by its own weight, and it is very simply to the effect that in the jugular veins these valves face downwards and not upwards. But these anatomical observations made on a cadaver are not sufficient of themselves to prove the competence of the valves in life or the direction of the flow of blood. Consequently, Harvey continues his work with various experiments using ligatures. And here again, Fabricius had suggested the possibility of this kind of investigation.

While wondering why other anatomists had not noticed the valves, Fabricius explains how he had seen them during a demonstration of the veins:

> When assistants pass a ligature round the limb preparatory to blood-letting, valves are quite obviously noticeable in the arms and legs of the living subject. And, indeed, at intervals along the course of the veins, certain knotty swellings are visible from the outside; they are caused by the valves. In some people, in fact, such as porters and peasants, they appear to swell up like varices. . . .

So also Harvey writes:

> let the arm of a living man be tied above the elbow as if it were to let blood. There will appear at intervals, especially in country people and those who are swollen veined, certain little nodes or swellings, not only where there is a branching, but likewise where there is none. These nodes are made by the valves.

Fabricius had found support for his belief that the valves slowed the impetus of the blood in the following experiment, and that Harvey saw it done in Padua around 1600 is not impossible. First, bind the arm as for blood-letting. Now,

> if one tries to exert pressure on the blood, or to push it along by rubbing from above downwards, one will clearly see it held up and delayed by the valves.

As long as pressure is exerted by stroking the blood downwards onto a valve in a vein that has not been emptied, its appearance is consistent with Fabricius's description and the vein will not be emptied. Harvey, however, realized that the only way to see whether blood leaked back beyond a valve was to empty the portion of the vein below it.

> If you press on the node with your finger or thumb and draw the blood downwards from that node or valve to the next, you will see that no blood can follow, the valve quite hindering it, and that the portion of the vein between the node and the finger you moved down is quite obliterated and yet full enough above the node or valve. Now, if you retain the blood so drove down and the vein emptied, and with the other hand press downwards in the direction of the valves, the upper part of the vein being full, you will find that by no means can it be forced or driven beyond the valve . . . [see p. 103, and Plate II].

By this experiment Harvey has proved that the valves of the veins are competent and that their action is similar to that of the valves of the heart, they permit the flow of blood in one direction only. At this point he says:

> Hence since anyone can try this experiment in many different places, it appears that the function of the valves in the veins is the same as that of the three sigmoid valves which are made in the orifice of the aorta and pulmonary artery, namely to shut so exactly that they do not allow the blood passing through to turn back again.

It now only remained for Harvey to prove that the blood in the veins flowed towards the heart and not away from it. This fact was already apparent from the preceding experiment, but only incidentally, and therefore Harvey proceeds to a further experiment on a ligated arm, illustrated in Plate II (Fig. 4). As before, put a finger firmly on the valve (at L) and empty the vein by stroking the blood upwards and above the next valve (at N), then

> taking away your finger (L), you will see it filled again from below. . . . And do this a thousand times in a little spacc.

From this experiment it plainly appears that the veins are the open and patent ways through which the blood returns to the heart. Fabricius, to describe the action of the valves, had likened them to dams in mill-streams contrived by engineers to keep back and accumulate water for the use of the machinery. But there is nothing in the human body which corresponds to this analogy, and Harvey with this same image of waterways in his mind likens the valves to flood-gates which open in one direction only and yield a passage to the accumulated pressure of the stream.

If Harvey did proceed in this way, his work on the structure and function of the valves of the heart, followed by that on the structure and function of the valves of the veins, would have brought him to a point at which he could have formulated the hypothesis that the blood circulates throughout the whole body. It seems, therefore, that there is no reason to doubt the truth of Boyle's report of his conversation with Harvey. The valves of the veins hold the key to the final solution of the problem and it was their manner of action which first 'induced him to think of' the circulation of the blood.

One further consideration was necessary, and that concerns the quantity of blood passing through the heart as compared with the total quantity contained in the whole body. To this question Harvey turns immediately after his experiment to show the direction of the flow of the blood in the veins:

> Now if you reckon the business, supposing how much blood in one compression has been moved upwards beyond the valve, and multiplying that by a thousand, you will find so much blood passed by this means through a little part of a vein in a short time, that I think you will find yourself perfectly convinced concerning the circulation of the blood and of its swift motion.

Though this may have convinced Harvey of the circulation, he did not let the matter rest there but, turning to the heart, proceeded to measure the content of the left ventricle, to estimate the quantity of blood extruded at each contraction and so to reach the conclusion that more blood passes through the heart than can be contained in the vessels of the whole body or produced *de novo* from the food eaten. And all this he sets out in Chapter 9 because it is the fundamental observation which underlies each aspect of his demonstrative proof. His own discovery that it is so may well have followed and not preceded his crucial demonstrations on the direction of blood flow in the veins which arose directly from his investigation of the venous valves.

Claude Bernard in his *Introduction à l'Étude de la Médecine Expérimentale* (Paris 1865) was of the opinion that an anticipative idea or an hypothesis was the necessary starting point for all experimental reasoning, and that without it no investigation of phenomena would be possible. Mere accumulation of data based on observation is not of itself sufficient to advance knowledge. If we admit the truth of this opinion, it follows that Harvey must have had an *a priori* or intuitive idea of the existence of the circulation of the blood before he began to consider which of his observations seemed crucial and before he designed experiments to test his initial hypothesis. 'I began to bethink myself whether it might not have a motion as it were in a circle. And this I afterwards found to be true.' Here is the time sequence that one would expect. Harvey's imaginative flash of intuitive understanding preceded the formulation of the hypothesis; the experiments to test it followed. What we cannot hope to know beyond all possible doubt is what sparked off this imaginative flash. Because Harvey said to Boyle that it was the valves in the veins, I have attempted to show from his writings that this is indeed a possible starting point for the train of events. Some critics reject Harvey's statement in favour of a belief that the idea arose from his realization of the quantity of blood, but I think their opinion is based on a mistranslation of Chapter 8 (see below, p. xlviii). Dr Pagel, who is immensely learned in the philosophical writings of the seventeenth century, has explained in his *Biological Ideas of William Harvey* why he thinks that it was the 'philosophy of circles' which first gave Harvey his idea, and he maintains that the hypothesis stems from his vision of analogy as revealed in the various analogies which Harvey cites after announcing his hypothesis in Chapter 8. Because my interpretation of Harvey's habits of thought as revealed in his writings is different from that of Dr Pagel, I find that I am not in agreement with him concerning the genesis of the idea. In all his writings Harvey insists again and again that the only way to achieve any knowledge of biological phenomena is to interrogate Nature herself. In *De motu cordis* he says:

> I do not profess either to learn or to teach anatomy from books or from the maxims of philosophers but from dissections and from the fabric of Nature.

And in Chapter 16, as I have already pointed out, he insists that it is his wish that above all else his hypothesis should be founded upon arguments based on anatomy. This same point of view we find expressed again and again in all Harvey's writings. He insists on the importance of personal observations many times repeated, the careful consideration of the phenomena observed, the weighing and assessing of their meaning.

> Wherefore it is that our judgement errs about phantasms and appearance comprised in our minds, unless sense give a right verdict, established upon frequent observations and unerring experience.

So he wrote in *De generatione*, and again in the same book he says:

> although it be a new and difficult way to find out the nature of things by the things themselves, rather than by the reading of books to take our knowledge from the opinions of Philosophers, yet must it needs be confessed that the former is a much more open way to the hidden secrets of natural philosophy and one which leads less into error.

From these and other passages like them, it seems to me that Harvey was not greatly influenced by opinions that he read in books, but greatly influenced by what he had himself observed. It is as if, at some stage in his career, he shut the door on his library in which he had spent many hours reading the authorities, and, taking with him his clarity of mind and his Aristotelian habits of thought, went out into his laboratory to confront the biological material which alone could provide him with the true answer to the many and diverse unsolved problems whose existence was all too plainly revealed in the books that he had read. He was a man of insatiable curiosity, but not of the kind of curiosity that speculates in the realms of the abstract. His contemplation was of the natural world around him: birds, animals, fish, reptiles, man, and all his questionings arose from what he had seen. The problems he set himself to solve were problems arising from his observations. The problem of the movement of the blood is a biological problem and its solution can only be expressed in biological terms. If he was indeed this manner of man for whom other men's opinions, even Aristotle's, were acceptable only when they had been confirmed by his own observation, then it is most unlikely that the imaginative act that created the hypothesis of the circulation arose from reflecting upon any abstract philosophical ideas, for his mind was entirely directed to the contemplation of the intractable problems of living biological material.

3. ON THE TEXT AND TRANSLATION OF CHAPTER 8

In the long and involved sentence not far from the beginning of this chapter, Harvey explains the reasons for which he had formed his hypothesis that the blood moves 'as it were in a circle'. But the text of the 1628 edition is not perfect and there is no general agreement among the subsequent editions as to how this passage ought to be read. As a result, there is, so far, no agreement either as to how it should be translated. It would, therefore, seem expedient to examine the Latin versions in an effort to isolate any errors and to decide how they should be corrected and the passage translated. The text of the sentence in question as printed in 1628 is as follows:

Sane cum copia quanta fuerat, tam ex vivorum, experimenti causa, dissectione, & arteriarum apertione, disquisitione multimoda; tum ex ventriculorum cordis, & vasorum ingredientium & egredientium Symmetria, & magnitudine, (cum natura nihil faciens frustra, tantam magnitudinem, proportionabiliter his vasibus frustra non tribuerit) tum ex concinno & diligenti valvularum & fibrarum artificio, reliquaque cordis fabrica, tum ex aliis multis saepius mecum & serio considerassem, & animo diutius evoluissem: quanta scilicet esset copia transmissi sanguinis, quam brevi tempore ea transmissio fieret, nec suppeditare ingesti alimenti succum potuisse animadverterim; quin venas inanitas, omnino exhaustas, & arterias, ex altera parte, nimia sanguinis intrusione, disruptas, haberemus, nisi sanguis aliquo ex arteriis denuo in venas remearet, & ad cordis dextrum ventriculum regrederetur.

Coepi egomet mecum cogitare, an motionem quandam quasi in circulo haberet, quam postea veram esse reperi, & sanguinem e corde . . . etc.

This text appears unaltered in the criticism of it by Parisianus in the editions of his work published in 1639 and 1647. Thereafter it was not reprinted until the edition of the text prepared by Albinus, published in Leyden in 1737. In 1766 it was used in the edition printed in London for the Royal College of Physicians (RCP). These two texts differ from each other and from the original in their punctuation. Compared with the 1628 version, that of 1737 reads as follows:

> *Sane cum copia quanta fuerat, tam . . . ventriculorum cordis,* [RCP omits comma] . . . *et magnitudine,* [RCP omits comma] (*cum natura* [RCP inserts comma] *nihil faciens . . . magnitudinem,* [RCP omits comma] . . . *tribuerit*) [RCP inserts comma] *tum ex . . . cordis fabrica,* [RCP uses semicolon] . . . *aliis multis* [RCP inserts comma] *saepius . . . considerassem,* [RCP omits comma] *et . . . evolvissem,* [RCP omits comma and inserts a bracket from *quanta* as far as *fieret*, with a semicolon after the closing bracket] *nec suppeditare . . . potuisse animadverterim;* [RCP uses comma] *quin . . . sanguinis intrusione,* [RCP omits comma] . . . *sanguis aliquo* [RCP changes to *aliqua*] . . . *regrederetur.* [RCP substitutes semicolon and runs on without making a new paragraph.]

That a certain dissatisfaction was felt with the 1628 text is clear from the fact that in the second edition of *De motu cordis* printed in Rotterdam in 1648, various changes were made. This text, compared with that of 1628, reads as follows:

> *Sane cum, copia quanta fuerit, saepius mecum et serio considerassem, tum ex vivorum (experiendi causa) dissectione . . . ex aliis multis: cumque animo diutius evolvissem, quanta scilicet . . . fieret; anne suppeditare ingesti alimenti succus eam posset: animadverti tandem, venas inanitas et omnino exhaustas, et arterias ex altera parte nimia sanguinis intrusione, disruptas fore; nisi sanguis aliquâ viâ ex arteriis . . . regrederetur.*

Coepi egomet mecum cogitare . . . etc.

Apart from slight differences of opinion as to whether it would be *his vasibus frustra* or *his vasis frustra*, this is the text contained in the editions printed in Rotterdam 1654, Rotterdam 1661, London 1661, Rotterdam 1671, Geneva 1685.

The only other variant which I have found is the version contained in the *Opera omnia* of Spigelius, printed at Amsterdam in 1645:

> *Sane* [beginning a new paragraph] *postquam tum ex vivorum experimenti causa dissectione, . . . et magnitudine, quam natura, nihil faciens frustra, his vasis proportionatam frustra non fecit; tum ex concinno . . . evolvissem, quanta scilicet esset copia transmissi sanguinis, et quam brevi tempore ea transmitteretur nec tamen suppeditari illam ab ingesti alimenti succo posse animadverterem, quin venas inanitas et omnino exhaustas, arterias vero nimia sanguinis intrusione disruptas haberemus, nisi sanguis aliquo . . . regrederetur: coepi egomet mecum cogitare* . . . etc.

From this comparison of the texts, it is clear that what has chiefly worried the editors is the lack in the 1628 text of a main verb in the sentence *Sane . . . regrederetur*. Two alternative possibilities have consequently been tried: (1) to run on the sentence to include the obviously main verb *coepi* and its dependent clauses; (2) to change *animadverterim* to *animadverti* thus making it into the main verb of the sentence and to adjust the grammar accordingly. Both alternatives present certain difficulties. In the absence of any evidence to the contrary, we must assume that the 1628 text, printer's errors apart, represents exactly what Harvey wrote in his manuscript. Though the 1648 emendation makes the meaning of the passage abundantly clear, it is not possible to account for the alterations as being merely corrections of printer's errors. If we keep the 1628 text and assume that its main verb is *coepi*, then we are faced with the problem of how to explain the construction and consequently the meaning of the passage *nec suppeditare . . . potuisse animadverterim; quin . . . haberemus* etc.

The first English translation which derives from the 1648 emended text reads as follows:

> and when I had a long time considered with my self how great abundance of blood was passed through, and in how short time that transmission was done, whether or no the juice of the nourishment which we receive could furnish this or no: at last I perceived that the veins should be quite emptied, and the arteries on the other side be burst with too much intrusion of blood, unless the blood did pass back again by some way out of the arteries into the veins, and return into the right ventricle of the heart.
>
> I began to bethink my self if it might not have a circular motion. . . .

The merit of this version is that is shows clearly the precise development in the various stages of the sentence: the quantity of blood, the speed of the transmission, the fact that the food eaten cannot supply enough, the remembered sight of empty veins and bursting arteries and the idea that to prevent this from occurring arteries and veins must somewhere, somehow communicate, and so to the final hypothesis of the circulation of the blood throughout the whole body.

Willis, who presumably worked from the RCP text of 1766, translated this passage as follows:

> . . . I frequently and seriously bethought me, and long revolved in my mind, what might be the quantity of blood which was transmitted, in how short a time its passage might be effected, and the like; and not finding it possible that this could be supplied by the juices of the ingested aliment without the veins on the one hand becoming drained, and the arteries on the other getting ruptured through the excessive charge of blood, unless the blood

> should somehow find its way from the arteries into the veins, and so return to the right side of the heart; I began to think whether there might not be a motion, as it were, in a circle.

When this translation is carefully considered, it becomes evident that it does not make very good sense for it puts together two separate and distinct considerations. The juice from the food which is eaten cannot supply the quantity of blood which passes through the heart. How can it then be said that this insufficient amount would produce empty veins and bursting arteries unless they communicated? The statement is surely a non-sequitur. The error has arisen from supposing that *quin . . . haberemus* etc. is dependent on *nec . . . animadverterim*, and this is grammatically impossible.

If we assume that *coepi* is the main verb of the sentence, then everything which precedes it is part of the protasis which reduced to its bare bones is a *cum* clause followed by two verbs in the pluperfect subjunctive: *cum . . . considerassem et . . . evoluissem . . . coepi. . . .* Over the tenses of these verbs there is complete agreement in all the versions of the text, and we may therefore assume that they were used by Harvey and that their use was deliberate. The presence of this pluperfect subjunctive is of capital importance for it implies that the sense of these verbs is not purely temporal, but either circumstantial or directly causal, that is to say that they relate to an action completed in the past and which is the cause of the action of the main verb of the sentence, in this case *coepi . . . cogitare.* (For example, the sentence *Quae cum cognosceret, Caesar in Italiam profectus est* means simply 'when Caesar learnt these things, he set out for Italy'; whereas *Quae cum cognovisset, Caesar in Italiam profectus est* indicates that it was the news which he had received which caused Caesar to proceed into Italy.) That being so, it is clear that the various things which Harvey enumerates were the cause of his asking himself whether the blood might not have a motion as it were in a circle. Though it is clear enough from the 1628 text that the two verbs *considerassem* and *animo . . . evoluissem* are not synonymous and that *considerassem* relates to what precedes it and *animo . . . evoluissem* to what follows, this distinction is made abundantly clear in the 1648 revision by the separation of these verbs in the sentence. This separation is, however, unnecessary. The primary meaning of *considerare* is 'to look at closely or attentively, to inspect', and that part of the sentence which depends directly on *considerassem* recounts the things which Harvey saw. *Animo evolvere*, literally 'to unfold in my mind', relates to what follows and which is summed up by *scilicet*, namely the large quantity of blood passing through and the short time this took, and both these dependent clauses are correctly framed as indirect questions with their main verbs in the imperfect subjunctive in accordance with the correct sequence of tenses.

We are now left with the problem of how to explain the construction of the remainder of the sentence. By changing *animadverterim* to *animadverti tandem* and omitting *quin*, the 1648 edition made the preceding clause depend on the more remote *evoluissem* and consequently had to introduce considerable alterations to change it from an accusative and infinitive construction into an indirect question in line with the two which precede it. But as the alteration involved goes far beyond the scope of merely correcting a printer's error, we are at liberty to question its validity, in spite of its obvious expediency. Harvey's Latin may not be Ciceronian in its elegance but it is both competent and grammatically correct as is to be expected from a man accustomed to use it in speech and writing and whose whole education from childhood had been conducted in it. I think we may, therefore, suppose that the offending perfect subjunctive *animadverterim* is merely a printer's error for *animadvertissem*. In this way the correct sequence of tenses would be maintained: *considerassem . . . animo evoluissem . . . animadvertissem*. At first sight the introductory *nec* is awkward, but the meaning is clear enough: 'when I had turned over in my mind these questions, namely, how great was the abundance of the blood that was passed through and in how short a time that transmission was done, and when I had perceived that the juice of the food that had been eaten could not suffice to supply it'.

At this point the 1628 text inserts a semicolon and though we are at liberty to disregard to some extent the original punctuation, alterations should only be made to it for valid reasons. The RCP text has replaced the heavy stop of the original by a comma, thereby implying that *quin . . . regrederetur* is dependent on what precedes. This has led Willis and others following him to suppose that *quin* is connected with the preceding negative, whereas in fact this is the use of *quin* that is often classified as corroborative and which occurs in such exclamations as *quin immo*, *quin etiam* and so forth. The passage does not depend either on what precedes or what follows. It is an exclamatory aside, almost a parenthesis, and the presence of the imperfect subjunctive is to be explained as potential.

The remaining problem concerns the beginning of this immensely long sentence: *Sane cum copia quanta fuerat. . . .* The only direct object of the verb *considerassem* is *copia* and perhaps *quanta fuerat* for everything else is preceded by *ex*. But clearly *copia quanta fuerat* is grammatically unacceptable for *copia* is not an accusative form. Recently the precise meaning to be attached to it has been called in question. Because the chapter bears the heading *De copia sanguinis*, etc., and because after *evoluissem* Harvey apparently repeats and expands the same words into the clause *quanta scilicet esset copia transmissi sanguinis*, it has been suggested that this extended meaning should also be applied to the first use of the word *copia*. This suggestion alters the meaning of the sentence without improving its grammar. It is true that the interpolation might conceivably fit with what immediately follows, and that, altering the order of the words and disregarding the grammar, we might read: *cum . . . considerassem . . . quanta copia sanguinis fuerat tam ex vivorum . . . dissectione*, and interpret: 'when I had considered how great abundance of blood there was both from the dissection of living animals for experiment's sake and from the opening of arteries. . . . '. But what then are we to make of the remainder of the clause that precedes *considerassem*? How can the symmetry and size of the ventricles and vessels of the heart, the symmetrical and careful structure of the valves and fibres and the fabric of the whole heart, relate to the quantity of the blood? Granted that the size of the vessels might be linked with it, I do not see how the precise structure of the valves of the heart can have anything whatsoever to do with the abundance of the blood. If this suggested interpolation were correct, then we would have an example of loose thinking and this is totally uncharacteristic of Harvey. None of the various editors from 1648 onwards suggested any such interpolation and it would seem to be both unnecessary and erroneous.

Undoubtedly, the main stumbling-block to the correct construing of this part of the protasis is the apparent use of *copia* without any noun in the genitive to define it.* We might, therefore, enquire whether this explanatory genitive is not in

* Such a use, however, is possible in Latin and it is clear enough from the context that the enumeration which follows makes up the *copia*. The first English translator understood it in this way: 'Truly, when I had often and seriously considered with my self, what great abundance there was, both by the dissection of living things', etc. But this translation underlines the difficulty of putting into English this unqualified *copia* when this one of its many meanings is attached to it. Willis, aware of the difficulty, translated it as 'When I surveyed my mass of evidence', thus making explicit what is implicit in the earlier rendering.

fact present in the text, disguised by some printer's error. Immediately we come upon the two words *disquisitione multimoda*. They seem to depend upon the preceding *ex* and they are intercalated, without the easing benefit of an *et*, between the first element of the enumeration, the references to experiments, and the second, the observations on structure. What is more, the word *disquisitio* is inappropriate in the context for it is a legal term, meaning strictly a judicial enquiry or investigation. The appropriate word is *inquisitio** for it relates to all the various investigations which Harvey has undertaken. That being so, it would be more appropriately placed before, rather than in the middle of the enumeration which it subsumes. I would, therefore, suggest that what Harvey wrote was *inquisitionum*, and that this is the noun in the genitive explaining *copia*. The printer has not only misread it but misplaced it, and because he printed it somewhere after an *ex*, has given it an ablative ending.

If we accept that the only possible direct object of the subordinate verb *considerassem* is *copia*, then it must be emended to *copiam* and the mistake explained as an error arising either from the printer's failure to notice the abbreviation sign by which Harvey customarily marked the accusative ending, or from his having been influenced by the preceding word *cum*, which he mistakenly assumed to be a preposition governing the ablative. (Most probably he had not read to the end of the sentence before he began to set it ip.) *Quanta* must be corrected to *quantam* as it agrees with *copiam*. The 1648 text changes *fuerat* to *fuerit*, but there is little to choose between them, provided it be understood that this is not an indirect question, in which case *esset* would be required, but a statement inserted almost as an exclamation in the main flow of the sentence. At this point, before the beginning of the enumeration, the

* In its legal sense, 'a searching for proofs'.

words *inquisitionum multimodam* should be inserted as the explanation of *copiam*.

Two further small errors need to be corrected. The enumeration cannot begin *tam* and be followed by *tum . . . tum*. The 1648 text changes *tam* to *tum*, but the idiom is more correctly *cum . . . tum . . . tum*. In the sentence in brackets, *cum natura . . . tribuerit*, the tense of the verb should be changed to the imperfect subjunctive *tribueret*, thus preserving the correct sequence of tenses and bringing it into line with a similar parenthetical statement in the second half of the protasis, *quin . . . haberemus*.

These emendations involve little more than the changing of a few letters and can all be accounted for as arising from the printer's inability to decipher Harvey's handwriting correctly. It was to avoid the occurrence of errors of this kind that Sir George Ent undertook to 'oversee the printing' of Harvey's *De generatione*.

> I made it my province to oversee the printing, and because the Author writes so obscure a hand (a thing as we may say, common to learned men), that scarce anyone but who has been accustomed to it can read it without difficulty, I used all diligence to provide against the errors of the compositor that might be occasioned thereby, a thing which I observe was not prevented in the printing of a small treatise of his not long since published.

No definitive Latin text of *De motu cordis* has yet been produced. A few obvious errors which affect the translation I have pointed out in the footnotes. My translation of the long and involved sentence in chapter 8 is based on the following emendation of the 1628 text:

> *Sane, cum copiam, quantam fuerat, inquisitionum multimodam, cum ex vivorum experimenti causa dissectione et arteriarum apertione, tum ex ventriculorum cordis et vasorum ingredientium et egredientium symmetria et magnitudine (cum*

natura nihil faciens frustra tantam magnitudinem proportionabiliter his vasibus frustra non tribueret), tum ex concinno et diligenti valvularum et fibrarum artificio reliquaque cordis fabrica, tum ex aliis multis, saepius mecum et serio considerassem, et animo diutius evoluissem quanta, scilicet, esset copia transmissi sanguinis, quam brevi tempore ea transmissio fieret, nec suppeditare ingesti alimenti succum potuisse animadvertissem, quin venas inanitas omnino exhaustas et arterias altera parte e nimis sanguinis intrusione disruptas haberemus, nisi sanguis aliquo ex arteriis denuo in venas remearet et ad cordis dextrum ventriculum regrederetur, coepi egomet cogitare an motionem quandam quasi in circulo haberet. . . .

4. ON THE ANATOMY OF THE HEART

In Harvey's time, the anatomy of the heart was described otherwise than it is today with the result that some of the terminology used by him can be misleading. Instead of thinking of the heart as consisting of two sides each containing two chambers, an atrium and a ventricle, for Harvey the heart was essentially a two-chambered organ consisting of the two ventricles only. In common with other anatomists, therefore, he at times uses the word 'heart' as a synonym for the ventricles or the ventricular part of the heart, and conversely the 'ventricles' as a synonym for the heart itself.

At this time also, the heart could still be regarded as a kind of diverticulum arising from the vena cava as it passed on its way carrying venous blood from the liver to the head. As there was no knowledge of the difference in the direction of the flow of blood in the superior and inferior parts of the vena cava, no differentiation was made between them. The right atrium was accordingly looked upon as the enlargement of the vena cava immediately above its opening into the heart, that is into the right ventricle, and not as part of the heart itself. In the same way, the left atrium was considered to be the expansion of the terminal portion of the pulmonary veins. From this it follows that the position of the two pairs of inlet and outlet valves which govern the direction of the flow of blood through the heart was described in different terms from those now currently in use. The two inlet valves are now known as the atrio-ventricular valves, and of these the tricuspid valve is at the inlet of the right ventricle and the mitral valve at the inlet of the left. Harvey and his contemporaries thought of them as being situated in the mouth of the vena cava and pulmonary

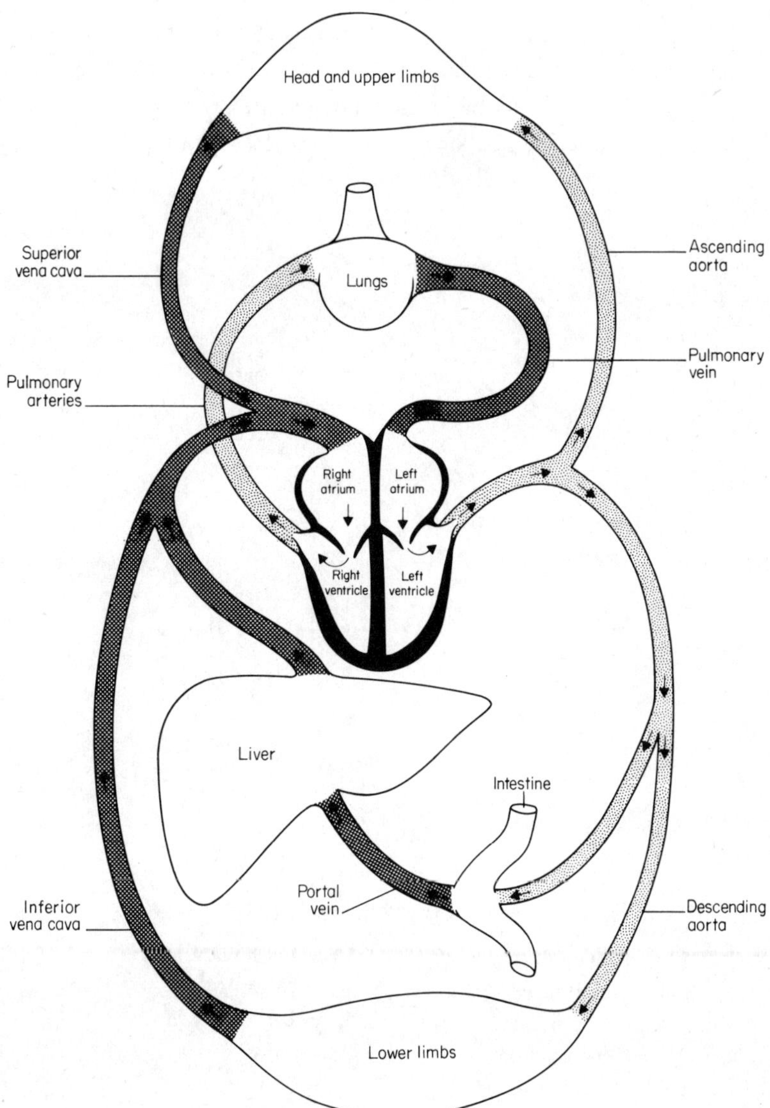

Figure 1. The general circulation. Arteries lightly shaded; veins heavily shaded.

vein respectively, and called them sigmoid, semilunar or tricuspid, and mitral or bicuspid. Similarly, the outlet valves, now known as the semilunar valves and differentiated as the pulmonary valve in the right ventricle and the aortic in the left, were described as situated at the base of the pulmonary

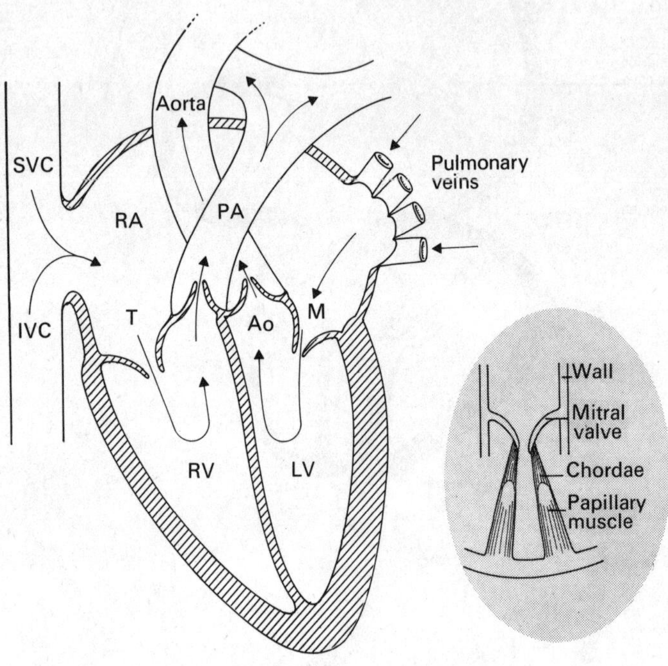

Figure 2. The heart showing blood flow and direction.

SVC = Superior vena cava
IVC = Inferior vena cava
RA = Right atrium
T = Tricuspid valve
RV = Right ventricle
PA = Pulmonary artery
M = Mitral valve
LV = Left ventricle
Ao = Aorta

Inset figure shows attachment of papillary muscles and chordae tendineae to the cusps of the mitral valve.

[Folkow B. and Neil E. (1971) *Circulation*, New York]

artery and aorta respectively and could all be called semilunar, sigmoid or tricuspid. They were differentiated only according to their position which it is therefore important to notice.

Since the right and left atria were regarded as being merely the expansions of the vena cava and pulmonary veins, it is not unusual to find the atria called by the names of these veins and this can be misleading, particularly with regard to the 'pulmonary vein'. (See particularly Chapter 6 where this use is frequent.) The openings of the four pulmonary veins into the left atrium were known from antiquity, but they were thought of as arising from one trunk, namely the left atrium. They could be mentioned in the singular or the plural depending on the way in which the writer was thinking of them, that is in the heart or distributed in the lungs.

The modern term atrium for each of the upper chambers of the heart has replaced the older auricle which is now applied to the auricular appendage. Harvey's use of the word auricle indicates that he also attributes this modern meaning to it.

The auricles or auricular appendages are merely the ear-shaped prolongations of the atrial walls and their contraction is simultaneous with that of the atria. It was their contraction as the heart passes into systole that Harvey observed and described as driving the blood into the ventricles, an observation which is now considered only partially true. While the heart is in diastole, the blood enters freely into both atria and both ventricles, for at that time the pressure on both sides of the inlet valves is equal and they are open. Towards the end of this process of spontaneous filling, the contraction of the atrial walls occurs and provokes the simultaneous contraction of both the ventricles. Only the quantity of blood remaining in the atria and in the auricles is actually driven on by their contraction. Yet Harvey's idea concerning the harmony and rhythm of the movement of the heart is basically correct. Clearly the heart can only function efficiently if the contraction and relaxation of the atria and the ventricles occurs in an ordered sequence. Harvey was obviously intrigued by the problem of how this movement was initiated and so was inclined to attribute it to

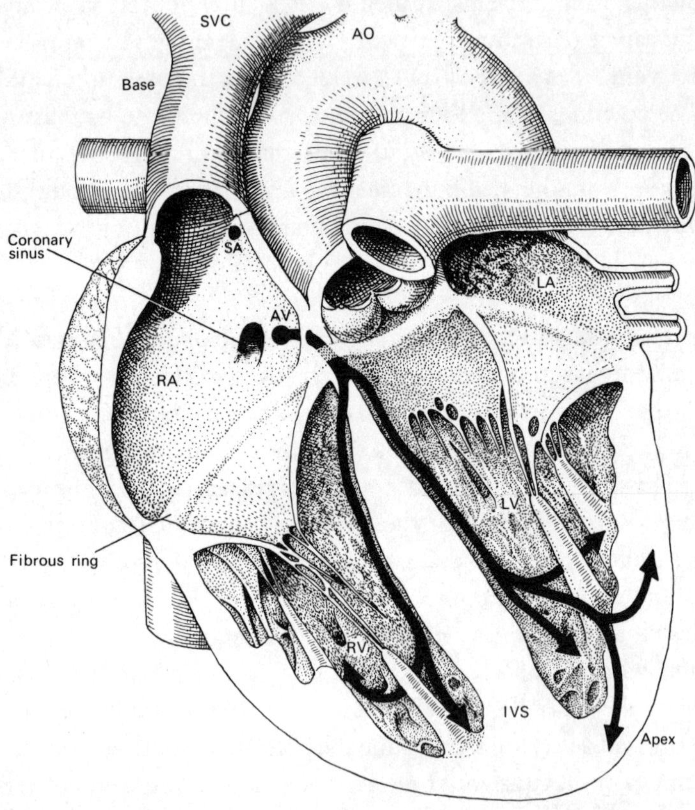

Figure 3. The arrangement of the specialized muscle in the heart: SA, sinu-atrial node; AV, atrioventricular node which leads to the bundle piercing the fibrous ring which separates the right atrium (RA) and the left atrium (LA) from the right ventricle (RV) and the left ventricle (LV); IVS, inter-ventricular septum; SVC, superior vena cava; AO, aorta.

the 'obscure palpitation' of the blood. He recognized fully the muscular nature of the heart but could not know that the peculiar characteristic of all the cardiac muscle fibres is their capacity to contract rhythmically and spontaneously, independently of the control of the nerves. Only within approximately the last hundred years has it been discovered that the wave of excitation starts from that part of the heart where the fibres have the highest rate of spontaneous contraction, that is from the sinu-atrial node, called the pacemaker, and from there spreads through the walls of the right and left atria, effecting their simultaneous contraction, and reaching the atrioventricular node. From there this wave of excitation passes through the bundle of specialized and very fast-acting muscle fibres (called after their discoverer who identified them in 1839, Purkinje fibres) which run down the septum of the heart and divide along its sides and are finally distributed into the walls of both ventricles causing them to contract simultaneously and immediately after the atria. When Harvey said, therefore, that the contraction of the heart is produced by the simultaneous contraction of all the muscle fibres of which it is composed, he was as right as he could be for his time. He describes these fibres as circular, but in the walls of the ventricles they are in fact arranged for the most part in spirals and form a vortex at the apex of the heart. His observation that the wall of the left ventricle is considerably thicker than the right in order to withstand the greater force of the contraction of the left ventricle as it projects the blood through the aorta into all the arteries of the whole body is also correct. The pressure of blood in the aorta (120 mmHg at peak) is about three times as great as that in the pulmonary artery.

Harvey's views on the foetal circulation are basically sound. (Fig. 5) The oxygenated blood passes from the placenta (which for Harvey served the office of the liver for the foetus)

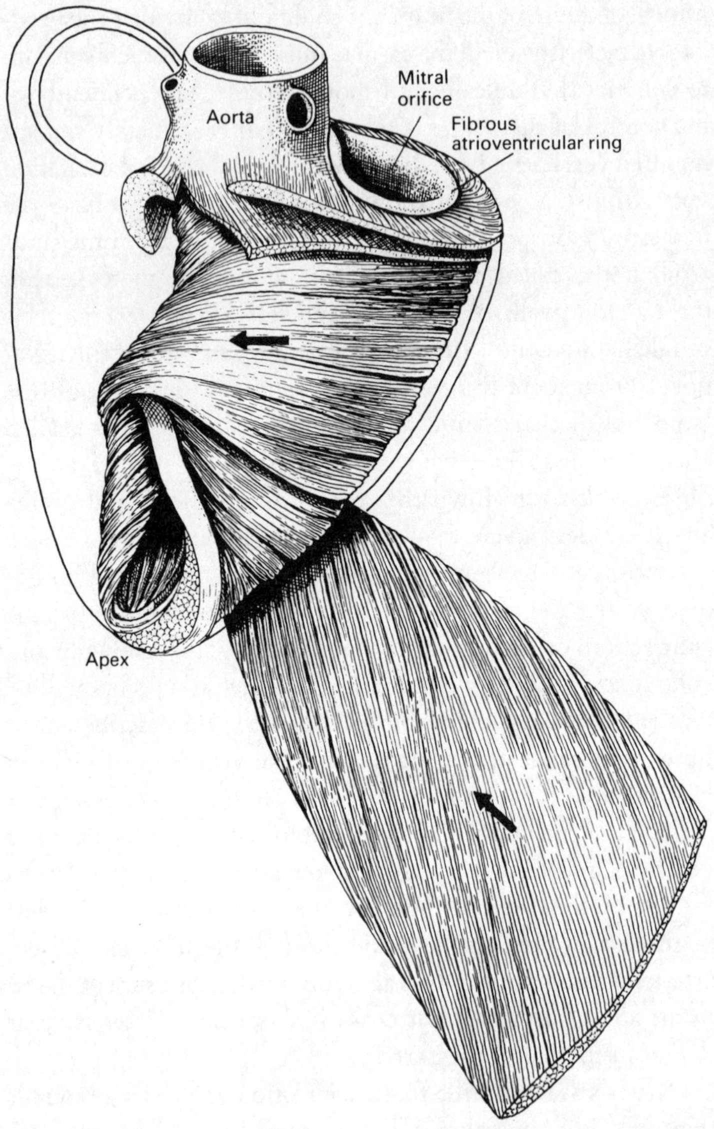

Figure 4. The bundles of fibres in the left ventricle of the human heart.

through the umbilical vein and eventually through the ductus venosus (DV) into the inferior vena cava (IVC) where it is mixed with de-oxygenated blood returning from the lower parts of the body. At the heart this blood stream is divided, by far the greater part of it going directly through the foramen ovale (FO) into the left atrium, the left ventricle and the aorta from which arises the blood supply to the upper half of the body (in the sheep by a single brachiocephalic artery, BCA). A small part of it, however, continues and meets the blood returning to the heart from the superior vena cava (SVC) with which it mingles and goes into the right ventricle (RV) and pulmonary artery. From the pulmonary artery most of this blood passes through the ductus arteriosus (DA) into the aorta. From the aorta some of it goes off to supply the arteries of the lower half of the body from which it is eventually collected into the inferior vena cava, while the remainder continues into the umbilical arteries to the placenta. A small part of the blood stream does not go through the ductus arteriosus but is carried by the other branches of the pulmonary artery to the foetal lungs where it completes the circuit by returning through the pulmonary veins to the left atrium.

Even though Harvey had discovered the circulation and knew precisely the direction of flow of the blood in the veins, no one should be surprised to find that he speaks of the smaller veins as arising from or being given off by the greater veins. The ancient terminology did not disappear from anatomical writings for many years after Harvey's discovery. Nor did the pulmonary vein and pulmonary artery immediately change their names but continued to be called the venous artery and the arterial vein.

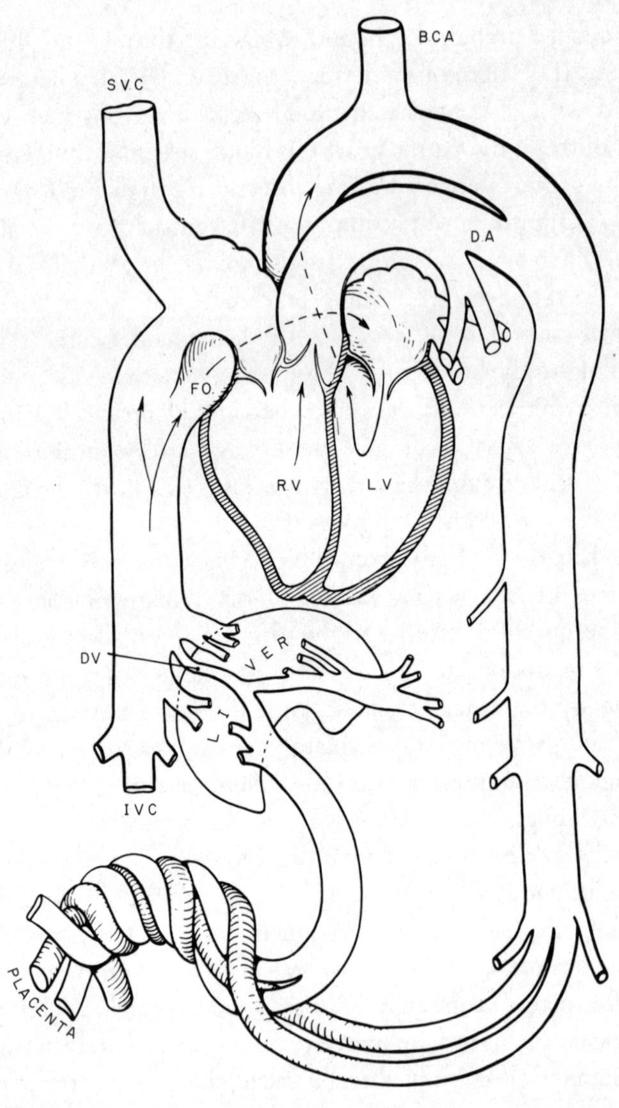

Figure 5. The foetal circulation of the sheep: IVC, inferior vena cava; SVC, superior vena cava; FO, foramen ovale; DA, ductus arteriosus; RV, right ventricle; LV, left ventricle; DV, ductus venosus; BCA, brachiocephalic artery.

BIBLIOGRAPHY

HARVEY, WILLIAM, *Exercitatio anatomica de motu cordis et sanguinis in animalibus*, Frankfurt 1628 [Keynes 1(a)]; Rotterdam 1648 [Keynes 1(b)].

Exercitatio de Circulatione Sanguinis, Cambridge 1649 [Keynes 30].

The Anatomical Exercises of Dr William Harvey, De motu cordis 1628: De circulatione Sanguinis 1649: The First English Text of 1653, ed. Geoffrey Keynes, London n.d. [1928] [Keynes 25].

Anatomical Exercitations Concerning the Generation of Living Creatures, London 1653 [Keynes 43].

Opera omnia, London 1766, 2 vols [Keynes 47].

The Works of William Harvey, M.D., trans. by Robert Willis, London, Sydenham Society, 1847 [Keynes 48].

De motu locali Animalium 1627, ed. and trans. Gweneth Whitteridge, Cambridge University Press, 1959.

The Anatomical Lectures of William Harvey: Prelectiones Anatomie Universalis; De Musculis, ed. and trans. Gweneth Whitteridge, Edinburgh and London, E. & S. Livingstone Ltd, 1964.

*

BAYON, H. P. William Harvey, physician and biologist, *Annals of Science*, III (1938), 59–118, 435–56; IV (1939), 65–106, 329–89.

CURTIS, JOHN G., *Harvey's Views on the Use of the Circulation of the Blood*, New York, Columbia University Press, 1915.

KEELE, K. D. *William Harvey: The Man, the Physician, and the Scientist*, London, Nelson, 1965.

KEYNES, SIR GEOFFREY, *The Life of William Harvey*, Oxford University Press, 1966.

The Bibliography of William Harvey, Cambridge University Press, 1928.

PAGEL, WALTER, *William Harvey's Biological Ideas*, Basel/New York, Karger, 1967.

William Harvey revisited, *History of Science*, VIII (1969), 1–31; IX (1970) 1–41. [Includes a copious bibliography.]

WHITTERIDGE, GWENETH, *William Harvey and the Circulation of the Blood*, London/New York, Macdonald & Elsevier, 1971.

ARISTOTLE. Quotations from Aristotle in the notes have been taken from the appropriate volume of the Oxford translation, ed. J. A. Smith and W. D. Ross.

FABRICIUS AB AQUAPENDENTE, HIERONYMUS, *De venarum ostiolis*, Padua, 1603.

GALEN. *Opera omnia*, 8 vols, Basel 1549.

MASSA, NICOLÒ, *Logica*, Venice, Bindoni & Pasini, 1550.

*

KNEALE, WILLIAM and MARTHA, *The Development of Logic*, Oxford University Press, 1962.

On disputations as still practised by the Dominican Order, see

GILBEY, THOMAS, O.P., *Barbara Celarent*, London, Longmans, 1949.

AN ANATOMICAL DISPUTATION CONCERNING THE MOVEMENT OF THE HEART AND BLOOD IN LIVING CREATURES

TO THE MOST
ILLUSTRIOUS AND INVINCIBLE MONARCH,

CHARLES,

KING OF GREAT BRITAIN, FRANCE
AND IRELAND,
DEFENDER OF THE FAITH

Most gracious King,

The Heart of all creatures is the foundation of their life, the Prince of all their parts, the sun of their microcosm, that on which all growth depends and from whence all strength and vigour flows. In like manner, the King is the foundation of his kingdom, the sun of his microcosm, the heart of his commonwealth, from whom all power flows and all mercy proceeds. Those things that are here written about the movement of the heart, I have been the more bold to offer to your Majesty (as is the custom of this age), seeing that almost all human affairs are after the pattern of man and most affairs of a King are after that of the heart. Therefore the knowledge of his own heart cannot be unprofitable to a King as being a divine example of his own actions (so have ever men been wont to compare great things with small). Most royal Sovereign, placed on the pinnacle of all things human, here at least you may contemplate at one and the same time the principle of man's body and the image of your kingly power. Accept, therefore, I most humbly entreat your most gracious Majesty, with your wonted kindliness and indulgence, these new things concerning the heart, you who are the new splendour of this age and indeed its whole heart, a prince abounding in virtue and grace, to whom we acknowledge our thanks to be due for any good that our England receives or any joy that our life holds.

Your sacred Majesty's
most devoted servant,
William Harvey

TO THE MOST EXCELLENT AND ACCOMPLISHED

DOCTOR ARGENT,[1]

PRESIDENT OF THE COLLEGE OF PHYSICIANS OF LONDON, HIS VERY DEAR FRIEND, AND TO THE REST OF THE DOCTORS AND PHYSICIANS, HIS MOST LOVING COLLEAGUES, GREETINGS

I have made known to you many times before, most worthy Doctors, my new opinion concerning the motion and use of the heart and the circulation of the blood in my anatomical lectures, but having now for nine years and more[2] confirmed it by ocular demonstrations in your sight, elucidated it with reasons and arguments, freed it from the objections of the most learned and skilful Anatomists, this opinion so often asked for by everyone and by some most earnestly requested, I have now brought forth into the light and view of everyone in this little book. Had it not passed through your hands, most worthy Doctors, I could scarcely hope that it would come forth safe and entire, for I can call on most of you as faithful witnesses of nearly all those observations from which I gather truth or refute error, you who saw my dissections and, in the ocular demonstrations of those things which I assert are plainly evident to the senses, were wont to stand by me and frankly agree with me. And since this book alone affirms that the blood goes forth along a new path and returns again, contrary to the way which has been accepted for so many hundreds of years by countless most famous and learned men and by them followed and well worn, I was greatly afraid to suffer this little book, otherwise complete some years ago,[3] either to come forth into the public view or go abroad,[4] lest it might seem too arrogant an action, unless I had first laid it before you, proved it by ocular demonstration, answered your doubts and objections and received the verdict of your most accomplished President in my favour.[5] I had, however, the utmost confidence that if I could sustain my argument in the presence of you and of our College, renowned for its so many and so greatly learned men, I then at last would have the less to fear from others, and the one comfort which for love of truth you granted me could then no less be hoped for from all the others who likewise are lovers of wisdom. For those who burn with the love of truth and wisdom are true philosophers, and never find themselves so wise, so full of wisdom, or so abundantly satisfied in their own knowledge but that they give place to truth from whomsoever or whensoever it may come. Nor are they so narrow-minded as to believe that any art or science was ever so absolutely or perfectly taught in all points by the Ancients, that there is nothing remaining to the industry and diligence of others, for there are indeed a great many who openly confess that the greatest part of those things which we do know, is the least of the things which we know not.[6] Neither do Philosophers suffer themselves to be so addicted to the slavery of any man's reports and precepts and so lose their own liberty that they no longer give credence to their own eyes, nor do they so swear allegiance to their Mistress Antiquity that they openly abandon or desert in the sight of all, Truth, their well-beloved. For as they think those men credulous and empty who at first sight accept and believe everything, so they take them for stupid and senseless who do not see things that are manifest to the senses, or acknowledge that day is day in the light of the noonday sun. They teach us to turn aside equally from the fables of poets and the follies of the rabble as from the suspensions of judgement of the Sceptics in matters philosophical. Again, all studious, good and honest men never suffer their minds to be so overwhelmed with the passions of indignation and envy that they will not patiently listen to what is spoken on behalf of truth or understand something which is truly demonstrated to them. They do not think it base to change their opinions if truth and openly demonstrated proof so persuade them, nor think it shameful to desert their errors, though they be never so ancient, seeing that they know full well that to err and be deceived is human, and that by chance are found out many things which any man may learn of another, an old man of a child, or a wise man of a fool.[7]

But, my loving Colleagues, I had no desire in this treatise to make a great volume and to parade my memory and labours and my readings by rehearsing and tossing about the names of authors and anatomical writers, of their works and their opinions, because I do not profess either to learn or to teach anatomy from books or from the maxims of philosophers, but from dissections and from the fabric of Nature herself.[8] And further, I do not endeavour nor think it fit to defraud any of the Ancients of the honour that is due to him, nor to provoke any of the moderns, nor do I think it seemly to contend and strive with those who have been excellent in anatomy and were my teachers. Moreover, I would not willingly lay an aspersion of falsehood upon anyone that is zealous for the truth, nor disgrace him with the stain of error.[9] I follow the truth alone and have expended all my time and trouble to this purpose only, to bring forth something that might be agreeable to good men, serviceable to learned men and profitable to the world of letters.

Farewell, most worthy Doctors,
and look favourably upon your Anatomist
WILLIAM HARVEY

NOTES

1 Dr John Argent, President of the College of Physicians 1625–34, was educated at Peterhouse, Cambridge, and admitted as a Fellow of the College of Physicians in 1597. He was present at Harvey's first examination by the College in May 1603 and in the following years they were frequently associated in the affairs of the College. Harvey refers to him several times in his *Anatomical Lectures* in connection with various cases and pathological findings. He also alludes to his presence at certain experiments. He practised medicine in London but retired to Broxbourne in Hertfordshire where he died in 1643. See the *Annals* of the College; *Munk's Roll*, vol. 1, London 1878; Sir George Clarke, *A History of the Royal College of Physicians of London*, vol. 1, Oxford 1964.

2 On 4 August 1615, Harvey was appointed Lumleian lecturer to the College of Physicians and continued to hold this lectureship until he resigned it in 1656. He gave his first public anatomy on 16, 17 and 18 April 1616. The notes for these lectures prove that at that time Harvey was not aware of the circulation of the blood. It is likely, therefore, that 'nine years and more' means precisely what it says and alludes to the period around 1618–19 as the time of the discovery. In the absence of any independent witness, we have no means of knowing when Harvey first demonstrated the circulation of the blood to the College either in private or in public. See my discussion, *WH and the Circulation*, chapter 5.

3 The year 1625 might be a reasonable guess.

4 Other translators have taken this as an allusion to the publication of *De motu cordis* (*DMC*) 'across the sea' in Frankfurt, but the Latin does not necessarily imply this. It merely repeats the meaning of the preceding verb and repetition of this kind is characteristic of Harvey's Latin style.

5 Harvey has here summed up the whole process of a medical disputation as conducted in the universities and as it underlies the structure of *DMC*.

6 Cf. Harvey's remarks to Dr George Ent in the Prefatory Epistle to *De generatione*:

> Although many things have been discovered by the learned men of former times, yet do I believe that far more still remains concealed in the darkness of uncharted Nature. Nay, I cannot forbear frequently to

wonder and sometimes to smile at those who persuade themselves that all things were so consummately and absolutely delivered by Aristotle and Galen, or by some other great name, that not so much as a single straw was left to be added.

7 Cf. Aubrey's remark on Harvey in his *Brief Lives*:

Ah! my old friend Dr Harvey—I knew him right well . . . had he been stiffe, starcht, and retired, as other formal Doctors are, he had known no more than they. From the meanest person, in some way, or other, the learnedst man may learn something.

Brief Lives, ed. O. Lawson Dick, London 1949, p. 130

8 Cf. *De Generatione*, Preface:

Although it be a new and difficult way to find out the nature of things by the things themselves, rather than by the reading of books to take our knowledge from the opinions of Philosophers, yet must it needs be confessed that the former is a much more open way to the hidden secrets of Natural Philosophy and one which leads less into error.

9 Cf. *Anat. Lect.*, p. 17:

Not to praise or dispraise, for all did well; as beholden to those who concluded erroneously for they lacked opportunity.

THE INTRODUCTORY DISCOURSE

in which it is shown that those things which have till now been written about the movement and use of the heart and arteries are by no means true

Seeing that we are thinking of the movement, pulse, action, use and usefulness of the heart and arteries, it is of considerable importance first to set forth those things which have been published by others, and to take notice of the things which have been commonly said and taught, so that what has been rightly spoken may be confirmed and what is false corrected in the light of anatomical dissection, personal experience many times repeated, and diligent and precise observation.

To this day, almost all anatomists, physicians and philosophers maintain with Galen that the use of the pulse is the same as that of respiration, and that they differ in one thing only, that the pulse depends upon the animal faculty, respiration on the vital, being alike in all other things in regard to their usefulness and their manner of motion. Hence they affirm, as does Hieronymus Fabricius ab Aquapendente in his book *De respiratione*, the most recently published work on the subject,[1] that because the pulse of the heart and arteries does not suffice to fan and refrigerate the heart, therefore the lungs were fashioned by Nature to surround the heart.[2] From this it is clear that whatsoever those in former times said about systole and diastole, and about the movement of the heart and arteries, they said it in relation to the lungs.

But since the movement and composition of the heart is different from that of the lungs, and the movement and composition of the arteries different from that of the chest, it is probable that different uses and benefits should arise from this, and that the pulse of the heart and its use, and likewise that of the arteries, should differ greatly from that of the chest and the lungs. For if the pulse and respiration serve the same uses, and if in diastole the arteries draw air into their cavities, as it is commonly said, and in systole send forth fuliginous vapours through the same pores of the flesh and skin, and, furthermore, if in the time between systole and diastole they contain air, at any one time they must contain either air or spirits or fuliginous vapours. What then should these writers reply to Galen, who wrote a book to show that blood was naturally contained in the arteries and nothing but blood, that is, neither spirits nor air, as we can easily conclude from the experiments and arguments in the same book.[3] And if in diastole the arteries are filled with air which they have taken in, in a greater pulse the greater will be the quantity of air entering. Therefore, if, when there is a great pulse, you immerse the whole body in a bath of water or oil, the pulse must needs be immediately either smaller or slower, seeing that it is difficult for the air to permeate into the arteries through the surrounding water of the bath, if not impossible. Moreover, since all the arteries, both the deep and the cutaneous, are distended at the same moment and with the same speed, how can the air pass through the skin, flesh and whole fabric of the body into its depths, as freely and swiftly as it can through the skin alone? And how do the arteries of the foetus draw air from the outside into their cavities through the mother's belly and the body of the womb? Or again, how do seals, whales and dolphins and the whole race of cetaceans and all fish in the bottom of the sea take in air through such a great mass of water, and give it out again by the swift pulse in the systole and diastole of their arteries? But to say that they sup up the air implanted in the water and return their fuliginous vapours into it, is not unlike a fiction.[4] And if in systole the arteries drive out the fuliginous vapours from their cavities through the pores of the flesh and of the skin, why not the spirits likewise, which they say are contained there too, since spirits are much finer than sooty fumes? And if the arteries in their systole and diastole take in air and give it out again, as the lungs do in respiration, why do they not do this when a wound is inflicted by the cutting of an artery? If the trachea is cut by a wound, it is as plain as the noonday sun that the air goes in and comes out in two contrary movements. But when an artery is cut, immediately the blood is driven out by force in one continuous movement, and it is obvious that air does not either go in or come out.[5] If the pulsations of the arteries refrigerate the parts of the body and ventilate them as the lungs do the heart itself, in what way is it commonly said that the arteries carry from the heart to each part of the body the blood crammed full of vital spirits that nourish the heat of the parts, awake this heat from slumber and restore it when it is spent? And how comes it that, if you tie the arteries, the parts are straightway not only numbed and cold, and look pale, but at last, cease to be nourished, which happens, according to Galen, because they are deprived of that heat which previously flowed out through all the parts from the heart, since it is clear from this that arteries carry heat to the parts rather than cooling and ventilation.[6] Besides, how can diastole at one and the same time draw spirits from the heart to warm the parts, and cold from outside the body? Further, although some affirm that the lungs, arteries and heart serve the same uses, yet they say that the heart is the laboratory of the spirits, that the arteries contain spirits and send them abroad, but, contrary to the opinion of Columbus,[7] they deny that the lungs make the spirits or retain them. And yet these same men affirm with Galen against Erasistratus that blood is contained in the arteries and not spirits.

These opinions seem so to war with one another and to refute each other, that all of them are not undeservedly suspect. That blood is contained in the arteries, and that the arteries carry blood only, is obvious from Galen's experiment (see his book, *That blood is contained in the arteries contrary to the opinion of Erasistratus*), from arteriotomy and from wounds, and likewise from the fact which Galen also affirms in very many places, that from a cut artery in the space of one half hour, the whole mass of blood will be emptied out of every part of the body in a great and forcible profusion. The experiment of Galen is this: 'If you bind the artery at both ends with a fine cord and slit it open lengthwise down the middle, you will find that what is contained in the artery between the two ligatures is nothing except blood.' And so he proves that it contains only blood. From this we can argue also in the same way: If, after tying and slitting the veins in the same manner, we find in the veins the same blood that is in the arteries (a thing which I have often tried myself in dead men and in living animals), we may conclude likewise by the same reasoning that the arteries contain the same blood as the veins, and nothing but the same blood.[8] Some, while they attempt to resolve this difficulty by affirming that the blood in the arteries is arterial blood, full of spirits, silently concede that it is the function of the arteries to carry the blood from the heart into the whole body, and that the arteries are full of blood, for blood that is full of spirits is nonetheless blood. And the blood, which is indeed blood and flows in the veins, no one denies that it is imbued with spirits. Even if the blood which flows in the arteries swells with greater store of spirits, yet it is to be reckoned that these spirits are inseparable from the blood, just like those which are in the veins, and that blood and spirit make one body (like whey and butter in milk, or heat and water in hot water), by which the arteries are filled and whose distribution from the heart the arteries perform, and this body is nothing other than blood.[9] Now if they say that this blood in the arteries is drawn out from the heart by means of the diastole of the arteries, they seem to imply that the arteries are filled with that particular blood in their distention, and not with the surrounding air as they said before. For if they say that they are filled by the ambient air, how and when do they receive the blood from the heart? If it happens in systole, then an impossibility befalls: either the arteries are filled whilst they are contracting, or they are filled without being distended. If, however, it is in diastole, they receive at the same moment blood, air, heat and cold all together and for two contrary uses, and this is indeed improbable. Further, when they affirm that the diastole of the heart and of the arteries is simultaneous and so likewise their systole, there is here another illogicality. For how can it be when two bodies so joined to each other as they are, are both distended at the same time, that one can draw anything from the other, or when they are both simultaneously contracted, that one can receive anything from the other?[10] Besides, it is perhaps impossible for any one body so to attract into itself any other body as to become distended, for to be distended is to suffer an action, unless it act like a sponge which, having been first constricted by some force from outside, sucks in as it returns to its natural constitution. And it is difficult to imagine that any such thing could happen in the arteries. But I believe that I can easily demonstrate, and have before now publicly demonstrated,[11] that the arteries are distended because they are filled like leathern bags or water-skins, and they are not filled because they are inflated like a pair of bellows with air. However, Galen's experiment in his book, *That blood is contained in the arteries*, shows the contrary, and it is as follows. Having exposed the artery, he cut it along its length and inserted into it a reed or hollow pipe, to the end that the blood could not leap out and that the wound should be stopped. 'So long', he says, 'as the artery is thus, the whole of it will beat, but as soon as you have passed a thread around over the artery and the pipe and tightened it like a noose and bound the coats of the artery to the reed, you will see that beyond the noose the artery no longer beats.' I have neither tried this experiment of Galen's, nor do I think it can be tried in a living body on account of the bursting forth from the arteries of the violently rushing blood, nor will the pipe close the wound without a ligature, and I am sure that the blood will rush out through the hollow pipe and beyond it.[12] However, by this experiment Galen seems to prove that the pulsific faculty flows from the heart through the coats of the arteries, and that the arteries, while they are being distended, are being filled by that pulsific faculty because they are being distended like a pair of bellows, and not distended because they are being filled like water-skins. But the contrary is obvious both in arteriotomy and in wounds, for the blood comes rushing out of the arteries, leaping violently, sometimes further, sometimes nearer, by fits and starts, but the leaping of it is always in the diastole of the artery and not in the systole. From this it plainly appears that the artery is distended by the impulsion of blood.[13] For while it is being distended, the artery cannot of itself throw out the blood with such force. Instead, according to those things which are commonly related about the use of arteries, it ought to attract air into itself through the wound.

And do not let the thickness of the arterial coats cozen us into believing that the pulsific faculty flows from the heart through these coats. For in some animals the arteries differ in nothing from the veins, and in the extremities of a man, in the tiny ramifications of the arteries, as in the brain, in the hand and so forth, nobody can distinguish arteries from veins by their coats, for they both have the same. Moreover, in an aneurysm proceeding from the incision or erosion of an artery, there is exactly the same pulsation as in any other artery, and yet it does not have the coat of an artery. In this the learned Riolan agrees with me in his [*Anthropographia*] Bk VII.[14]

And do not let any man believe that the use of the pulse and of respiration is the same, simply because he can see that the pulses become more frequent, larger and swifter, just as respiration does and from the same causes, namely from running, anger, a hot bath, or as Galen said, any other thing which heats. For not only does that experiment show the contrary (which Galen endeavours to explain away), that by excessive repletion the pulses are greater, and the breaths smaller, but likewise, in children the pulsations are frequent while at the same time the respiration is slow. Likewise in fear, care and anxiety of mind, even in some fevers, the pulses are swift and frequent, but respiration is slower.

These and other difficulties like them follow from the opinions that have been set down concerning the pulse and the use of the arteries. Perhaps those things which have been asserted concerning the use and pulse of the heart are no less entangled with very many and inextricable difficulties. Men commonly assert that the heart is the fountain and laboratory of the vital spirits by which it gives life to all the parts, and yet they deny that the right ventricle makes the spirits and say that it only provides nourishment for the lungs. And hence they say that fish have no right ventricle of the heart,[15] and indeed that it is altogether wanting in those creatures that have no lungs, and that the right ventricle of the heart was made for the sake of the lungs.

1. Why then, I ask you, since the constitution of both

ventricles is almost the same in that they have the same construction of fibres, muscular bands, valves, vessels and auricles, and in dissection are both found full of blood that is in both alike blackish and alike lumpy, why, I say, should we think that they were appointed to such diverse uses, seeing that their action, motion and pulsation is the same in both? If the three tricuspid valves at the entrance into the right ventricle are a hindrance to the return of blood into the vena cava, and if the three semilunar valves in the orifice of the pulmonary artery are made to impede the regress of blood, why, since things are similarly arranged, should we deny that the valves in the left ventricle were made likewise to hinder the egress and regress of blood?

2. Why, since the valves are almost identical in size, shape and position in the left ventricle as in the right, why do they say that here in the left ventricle they hinder the egress and regress of spirits, and in the right that of blood? It seems unlikely that the same organ can possibly be adapted to hinder the motion of blood and of spirits alike.

3. And why, since the passages and vessels correspond to one another in point of size, that is the pulmonary artery and the pulmonary vein, why should the one be destined to a particular use, namely to nourish the lungs, and the other to a general use?

4. And, as Realdus Columbus observed, how is it provable that so much blood is needed to nourish the lungs, for indeed this vessel, the pulmonary artery, is bigger in size than both the branches of the descending vena cava that supply the femoral veins?[16]

5. And lastly I ask why, since the lungs are so near and the vessel so great and the lungs in continual motion, why is there need for a pulse in the right ventricle, and why is it that Nature, for the sake of nourishing the lungs, found it necessary to add a second ventricle to the heart?

When they say that the left ventricle draws material from the lungs and from the right ventricle of the heart, that is to say air and blood, to prepare the spirits, and likewise distributes the spiritous blood into the aorta, and that from this left ventricle fuliginous vapours, on the one hand, are sent back through the pulmonary vein into the lungs, and, on the other, spirits into the aorta, what is it that makes the separation between them, and how is it that spirits and fuliginous vapours pass this way and that without mingling or confusion? If the mitral valves do not hinder the return of fuliginous vapours to the lungs, how can they hinder the return of air? And how do the tricuspid semilunar valves prevent the return of spirits from the aorta in the subsequent diastole of the heart? And by what manner of way do they say that the spiritous blood is distributed from the left ventricle through the pulmonary vein into the lungs, and the mitral valves not hinder it the while? seeing that they affirm that through this same vessel the air enters into the left ventricle from the lungs, the air to whose return they would have these mitral valves to be a hindrance. Good God! How do the mitral valves impede the return of air and not of blood?[17]

Furthermore, why, seeing they have destined the pulmonary artery, a large broad vessel made with the coat of an artery, to one use only and that a special one, namely to nourish the lungs, why do they assert that the pulmonary vein, which is scarcely of the same size and has the loose, soft coat of a vein, was made for several uses, three or four? For they will have the air pass through it out of the lungs into the left ventricle, and they will likewise have the fuliginous vapours return through it out of the heart into the lungs, and they will have the spiritous part of the blood to be distributed through it from the heart into the lungs to warm them back to life.

If they will have fuliginous vapours to be sent out from the heart, and air into the heart through the same pipe, then I must reply that Nature is not wont to construct one vessel and one way for such opposite motions and contrary uses, nor is such a thing ever to be seen anywhere.

If they maintain that fumes and air go out and return by this way, as they do through the bronchi of the lungs, why, when we cut out or open up the pulmonary vein, can we never find either air or fumes in the dissection, and how comes it that we always see this pulmonary vein crammed full of thick blood and never of air, whereas we find air remaining in the lungs?[18]

If anyone should try the experiment of Galen and cut the wind-pipe of a dog, being yet alive, and forcibly fill the lungs with air with a pair of bellows, and when they are filled, ligate them tightly, then, having quickly slit open the chest, he will find great store of air in the lungs right to their outermost coat, but none at all either in the pulmonary vein or in the left ventricle of the heart. But if in a living dog, either the heart drew air in from the lungs, or the lungs drove it into the heart, so much the more should they do so in this experiment. Indeed, in an anatomical dissection also, when the lungs of the cadaver have been inflated, who doubts but that the air would immediately enter these parts, did any passages for it exist? But they make so important this use of the pulmonary vein for the conveying of air from the lungs to the heart that Hieronymus Fabricius ab Aquapendente maintains that the lungs were made for the sake of this vessel, and that it is the chiefest part of the lungs.[19]

But, I beseech you, if the pulmonary vein were made for the conveyance of air, why does it have the constitution of a vein? Nature had need rather of pipes and those annular ones, like those in the bronchi, so that they might always stay open and not collapse, and so that they might remain altogether empty of blood lest its wetness should hinder the passage of air, a thing which is plain enough when we breathe with a whistling and a rumbling noise when the lungs are in trouble, either stuffed with phlegm from the bronchi, or at least laden with a little.

Still less to be tolerated is that opinion which, supposing that two kinds of material, air and blood, are necessary for the making of the vital spirits, does assert that the blood seeps through hidden pores in the septum of the heart, out of the right ventricle into the left, and that the air is drawn in through the great vessel, the pulmonary vein, out of the lungs, and therefore in the septum of the heart there are many pores adapted for the exuding of blood. But, in God's truth, there are no such pores, nor can any be demonstrated.[20]

For the substance of the septum of the heart is thicker and more compact than any part of the body, except the bones and sinews. But if there were holes, how were it possible, since both the ventricles are distended and dilated at the same time, for the one to draw anything from the other, or the left draw blood from the right? And why should I not rather believe that the right ventricle draws spirits from the left than that the left, through the same holes, draws blood from the right?[21] But it is truly wonderful and absurd that, at the same instant, blood should be more conveniently drawn through hidden and obscure passages and air through very open ones. And why, I ask you, do they have recourse to hidden, invisible, uncertain and obscure pores for the passage of the blood into the left ventricle, when there is such an open way through the pulmonary vein? Truly, it is a wonder to me that they should make, or rather invent, a way through the septum of the heart, which is gross, thick, hard and most compact, than through the patent pulmonary vein, or else through the substance of the lungs, thin, loose, most soft and spongious. Besides, if the blood could pass through the substance of the septum or be imbibed from the ventricles, what need were there of the branches of the coronary vein and artery that are distributed for the nourishment of the septum itself? And this is most worthy of note: if, in the foetus, where all things are thinner and softer, Nature was forced to bring the blood through the foramen ovale, out of the vena cava through the pulmonary vein into the left ventricle, how can it be likely that in a grown man, it should seep so conveniently and with no trouble through the septum of the heart now made more dense with age?[22]

Andreas Laurentius, in Bk IX, cap. 11, Quaes. 12, backed by the authority of Galen, *De locis affectis*, Bk VI, cap. 7, and the experience of Hollerius, asserts and proves that in empyema, watery fluids and pus from the cavity of the chest being absorbed into the pulmonary vein, can pass through the left ventricle of the heart and the arteries and be expelled with the urine or the faeces.[23] In confirmation of this, he relates the case of a certain melancholic who often suffered from fainting and who was freed from a paroxysm by the emission of turbid, stinking, acrid urine. When he at last died from this kind of disease and his body was opened, no such substance as he pissed appeared anywhere either in the bladder or in the kidneys, but there was a great deal of it in the left ventricle of the heart and in the cavity of the chest. And so Laurentius boasts that he had foretold some such cause for diseases of this kind. But I cannot choose but wonder, since he had guessed and foretold that heterogeneous matter could be evacuated by this same passage, that he either could not or would not see or affirm, that through the same ways the blood could be conveniently and according to Nature, brought out from the lungs into the left ventricle.

Therefore, from these and many other such things as these, it is clear that those things which were spoken before by former writers concerning the motion and use of the heart and the arteries, seem either improbable or obscure or impossible, if one consider them rather carefully, and therefore it will be profitable to search more deeply into the business and to contemplate the motions of the heart and arteries not only in man, but also in all other creatures that have a heart, and likewise by the frequent dissections of living things and by many anatomies of the dead, to discern and search out the truth.

NOTES

1 Hieronymus Fabricius ab Aquapendente, the most important of Harvey's teachers in Padua and the one to whom he acknowledges a special debt in regard to his work on animal generation: 'In chief of all the Ancients, I follow Aristotle, and of later writers Hieronymus Fabricius of Aquapendente.' Both shared the capacity for minute, patient and careful observation and the subjects which had chiefly interested Fabricius were those to which Harvey also devoted most attention. Fabricius was appointed lecturer in surgery in the University of Padua in 1565 and, in due course, such was the respect in which he was held by his contemporaries, the chair of anatomy was created for him. He died in 1619. For details of his life and work, see Howard B. Adelmann, *The Embryological Treatises of Hieronymus Fabricius of Aquapendente*, New York 1942.

His *Tractatus de respiratione et eius instrumentis* was first published in Padua in 1615 and Harvey alludes to it on several occasions in his *Anatomical Lectures* of 1616 and notes his disagreements with Fabricius's conclusions. Harvey describes it here as *nuperrime edito*, which most translators have taken to mean 'recently' or 'very recently published'. As it is possible that Harvey did not write this chapter within five years of the date of publication of the treatise, it could be that he is alluding to it as the most recently published work on the subject, unless he is copying from some earlier notes.

2 This error which is the basis on which rests all the confusion of thought concerning the function of the heart and lungs which prevailed, as Harvey says, until his own day, derives from the opinions of the Greeks, notably Aristotle and Galen. It won easy acceptance from the fact that in man the lungs surround the heart (Aristotle, *De partibus animalium* (*P.A.*) 665a 15: 'where the heart is, there and surrounding it is the lung'), though Aristotle was well aware that this was not true for all animals (*Ibid.* 669a 21). Though

Galen's opinions on the subject were not entirely consistent, he does say in his treatise on respiration: 'That some portion of the air is drawn into the heart in its diastole and fills the vacuum which is produced, is sufficiently shown by the very magnitude of the dilatation.' In *De usu partium*, VI, 2 and elsewhere, Galen expresses the view that the respiration of animals arises from the heart, that the heart needs air to cool its boiling heat and this is effected by the drawing in of cold air in inspiration and the driving out in expiration of the sooty vapours which have been formed in the heart.

3 Empedocles, who flourished in the first half of the fifth century B.C., seems to have been the earliest philosopher to hold the opinion that the purpose of respiration was to control the natural heat of the body, and that respiration and the movement of the blood were one process. He believed that air, together with the *pneuma*, or soul, that was present in it, entered the body at the moment of inspiration not only through the mouth and nostrils, but also through the pores of the skin and the ends of the arteries underlying them, and that the alternate movements of inspiration and expiration were caused by the heartbeat. When the thin, effete blood, whose supplies of nourishment had been used up by the parts, returned to the heart at the time of its diastole, it left the extremities of the vessels empty and the air rushed into them through the pores of the skin, to be driven out again when the reinvigorated blood returned from the heart at the moment of its systole. So the systole of the heart was equated with expiration and the diastole with inspiration. For want of a better word, this process is often referred to as 'transpiration' and it is the notion which underlies Plato's discussion of the problem in the *Timaeus*. Later, it came to be thought that the air, and with it the spirit, went as far as the heart in its diastole and from there was distributed with the blood to the various parts of the body. At this time there was no clear distinction between veins and arteries. That came later and with it the belief that the arteries, as their name implies, were empty of any content save air.

When Galen proved conclusively in his attack on Erasistratus that the arteries contained blood at all times and nothing but blood, he overturned with a true observation a neat piece of theoretical ratiocination. But the truth did not win for itself general acceptance by all who came after him. For those who theorized but did not look, it was clearly inacceptable; and those who looked were faced with the dilemma: either to believe their eyes and reject time-honoured theory, or cling to the theory in spite of their own observation. So compromises came to exist and even anatomists of the calibre of Vesalius made do with a compromise. Among Harvey's contemporaries there were still those who were content with muddled thinking. And so Harvey is here concerned to spell out the contradictions that are inherent in the theory and the observed facts which refute it.

The 'fuliginous vapous' or sooty fumes, to which he alludes, are the residue of concoction in the left ventricle of the air, brought in through the pulmonary vein, with the blood which has seeped through the interventricular pores. It was axiomatic to their way of thinking that every concoction wrought by the agency of the innate heat of the body, left a residue which was subsequently purged from the body. The concoction of air and blood produced vital spirit, and its residue, in the shape of sooty fumes, was purged by being sent out of the left ventricle back through the pulmonary vein into the lungs and so breathed out of the body.

4 Several generations after Harvey's time it was, of course, found to be true. Aristotle was of the opinion that, because there was no air in the water, fish must take it in with their food (*De spiritu*, chapter 2, 482a 23). Whether Harvey thought this or whether he may indeed have had a suspicion that somehow air was contained in the water and that this was essential for the continued existence of fish is hard to say from the following remark in his *Anatomical Lectures*: 'There are certain animals like the cetaceans which, although they spend a long time under water, yet if they do not sometimes breathe, they die, porpoises, and so also fish in a frosty pond' (p. 297). But the reference to fish in a frosty pond is not original and the respiration of cetaceans was discussed by Aristotle in *P.A.* 16–21, and *De respiratione*, 476b 25. Harvey's remarks on the subject in the *Lectures* derive from this source.

5 *Anat. Lect.*, p. 269: '. . . when the arteries are wounded, the blood spurts forth and the air does not enter'.

6 *Ibid.*, p. 269: 'when an artery is obstructed, the part becomes livid and very cold'.

7 See below, p. 67.

8 Harvey's insistence on the fact that the blood in the arteries was the same

as the blood in the veins is well known. Venous blood was held to be imbued with natural spirits and so capable of bringing nutriment to the parts. No writer ever seems to have thought of these spirits as in any way separable from blood.

9 *Anat. Lect.*, p. 293: 'spirit and blood are one thing, like whey and cream in milk, and following the definition of Aristotle [*P.A.* 649b 22–7] the very word blood connotes the presence of heat as an accident in a substance which is essentially cold, just as does the word hot-water which implies as it were both steam and flame, the latter being the act by which the former is made actual, and light is the result of the act of something which gives light. As light is to the candle, so is the spirit to the blood and it has actuality in being made, like flames from fire, existing in continual generation and flux.' See also *De motu locali animalium*, p. 103: 'Blood and spirit are one thing'; and Harvey's long discussion of the problem in *De generatione*, Ex. LI and LII.

10 *Anat. Lect.*, p. 271: 'if the arteries and ventricles move simultaneously, then the movement does not arise from the heart, because if it were so then the heart would have two movements at the same time'.

11 This would seem to allude to Harvey's demonstrations and proof, in the course of his *Anatomical Lectures* before the College of Physicians, of the nature and timing of the systole and diastole of the heart, in which he upheld the belief of Realdus Columbus that it was the reverse of what was commonly supposed. See *Anat. Lect.*, pp. 267–9.

12 *Anat. Lect.*, p. 269: 'And Galen's argument from the reed or "pervious tube" is impossible.' Two accounts of this experiment occur in Galen's writings, one in his treatise against Erasistratus to which Harvey refers, and the other in *On Anatomical Procedures*, VII, 16, pp. 199–200. In 1649, in his Second Letter to Riolan, Harvey describes how he had himself performed the experiment and uses it to prove the reverse of Galen's contention: 'it supplies nothing in support of the opinion that the coats of the vessel are the cause of the pulse; it much rather proclaims that this is owing to the impulse of the blood' (ed. Willis, pp. 110–11). Galen's anomalous findings have never been satisfactorily explained.

13 *Anat. Lect.*, p. 269: 'the arterial pulse does not come from some innate pulsative faculty, but from the thrusting forth of the blood by the heart'.

14 Riolan, *Anthropographia*, Paris 1618: 'Nor is the pulsific force inherent in the arteries themselves, but it derives from the arterial blood which the heart pours out into the arteries, for in a raised and swollen aneurysm proceeding from the incision of an artery, the pulsation of the swelling comes from nowhere but from the arterial blood.' This was not the edition used by Harvey, but the quotation will be found here in Bk VI, in the chapter at the end of the volume entitled 'Viventis animalis observationes anatomicae', p. 85.

15 In *Anat. Lect.*, p. 255, Harvey had said that in fish 'the left ventricle is wanting and not, as is commonly supposed, the right'.

16 Columbus, *De re anatomica* (Venice 1559), VII, p. 178: 'This pulmonary artery of which we are speaking is of considerable size, truly much bigger than were necessary if its only use were to convey blood from the heart to the lungs over so short a distance.' And, XI, 2, p. 223: 'This pulmonary artery is so large that it may well be that it brings blood for some additional purpose, besides the nourishing of the lungs.' As the pulmonary artery was classed among the veins, its comparison with the size of the common iliac veins should cause no surprise.

17 One of the very few anatomists before Harvey's time to try to sort out the muddle inherent in all these diverse and contradictory notions was Berengario da Carpi in his *Commentaria cum amplissimis additionibus super anatomiam Mundini*, published in Bologna in 1521. Having quoted the various opinions of the ancient authorities, Galen and Avicenna, and those of the more modern anatomists in such a way as to underline their inconsistencies and contradictions, Berengario concludes in despair: 'Whatsoever may be the truth in these matters, I do not want to make any decision about them, for they are all much more imaginary than real' (p. cclv).

18 Typical of the muddled thinking about the pulmonary vein is the account of it given by Vesalius in his *De humani corporis fabrica* (Basel 1543). In one place he says of it that

ceaselessly and tirelessly it has to take into itself air from the branches

> of the arteria aspera, indeed from the lung itself, and with its own movements corresponding to the rhythms of the heart, to drive back again into the branches of the arteria aspera the fuliginous residues. . . . (Bk VI, ch. 15, p. 598).

But because he knows from his experience in dissection that it does contain blood, he says of it elsewhere that like all the arteries in the body, except the arteria aspera, 'it is filled with fine and spiritous blood and supplies suitable blood to the lung'.

19 The statement derives directly from Fabricius's treatise *De respiratione et eius instrumentis*. As in most of his works, Fabricius sticks closely to Galenic theory. He is convinced that the chief use of the pulmonary vein, for whose sake the lungs exist, is to draw in air to cool the heat in the left ventricle of the heart and spends much time in discussing how this affects the generation of vital spirits and the innate heat of the heart.

20 Though Berengario da Carpi does not utterly deny the existence of pores in the interventricular septum of the heart, he is not altogether happy about them and says that in a cadaver they are more imaginary than visible (*Commentaria*, p. cccxlvii^v). The first person to state that they did not exist was Niccolò Massa in his *Liber introductorius*, published in Venice in 1536: 'this middle wall is of a closely-textured, hard substance and without any cavity'.

In the *Anatomical Lectures*, Harvey had said: 'Some think that the blood passes across through the septum of the heart, the dividing wall, . . . and therefore they say that the septum is porous which X [it is not]. Bauhin, however, says that the vents were conspicuous in the heart of an ox which had been boiled.' Bauhin's unfortunate finding in what was certainly an over-boiled ox heart proved a stumbling block to many, including himself, for although he was ready to believe that the pores did not exist in 1590 in his *De humani corporis fabrica Libri IIII* (p. 258), in his *Theatrum anatomicum* of 1605, he said that: 'These pores are very conspicuous in the heart of an ox after it has been boiled for a long time' (p. 422). As late as 1635, Pierre Gassendi described in his *De septo cordis pervio observatio*, how he had seen a hole in the septum demonstrated by a surgeon at an anatomy at Aix. But Caspar Hofmann in his commentary on Galen's *De usu partium* (Frankfurt 1625, p. 112) explains that this was a trick.

21 Vesalius's ambivalence on the subject of interventricular pores is well

known. Though he could not find them, he never actually denied their existence and continued to allow some blood to seep through from the right ventricle to the left. What is more, he seems to have been the only anatomist to have thought that the right ventricle actually did draw blood from the left through these same invisible pores. When describing the action of the heart in Bk VI, cap. 15, he says: 'every time it dilates itself, the heart attracts into the right ventricle blood from the vena cava, from the lung and from the left ventricle of the heart' (p. 596). There is no other suggestion, however, in his account to prove that he really believed in this two-way passage. His statement that the right ventricle attracts blood from the lung is to be explained by the fact that he had no understanding whatsoever of the competence of the valves of the heart.

The possibility of the existence of this two-way flow was raised also by Bauhin in 1590 in his *De corporis humani fabrica Libri IIII*, p. 258: 'If the venous blood can flow so easily through into the left ventricle, why, having been made spiritous and warmer and finer, does it not flow back into the right ventricle through the selfsame pores in the septum and make a confusion in the order of things?'

22 Harvey returns to this discussion below in Chapters 6 and 7. In this sentence, Harvey is using 'pulmonary vein' with the meaning of the left atrium.

23 Cf. *Anat. Lect.*, p. 169: 'Into the renal vein is thrown a branch, sometimes two, from the azygos vein, hence matter from the chest is voided through the urine.' According to Laurentius, *Historia anatomica* (Frankfurt 1600), VI, 23, p. 253, this was the opinion of younger writers, but he himself preferred the Galenic view. In the same work, Bk IX, Q. 12, pp. 361–3, he examines and defends the Galenic theory that pus is absorbed from the chest into the pulmonary vein and goes through the left ventricle of the heart and the arteries into the kidneys. Among the 'younger writers' was Caspar Bauhin whose textbook formed the basis for Harvey's Anatomical Lectures. Bauhin gives both views and in a footnote, the references to Laurentius, Galen and Holler. That Harvey in 1616 seems to accept the passage of pus through the veins may be taken as an indication that he had not then any knowledge of the circulation of the blood.

CHAPTER 1

The causes which moved the Author to write

When I first applied my mind to observation from the many dissections of living creatures as they came to hand, that by that means I might find out the use and benefits of the motion of the heart through actual inspection with my own eyes, and not from books and the writings of other men, I straightway found it a thing hard to be obtained and full of difficulty, so that I almost believed with Fracastorius that the motion of the heart was known to God alone. For I could rightly distinguish neither how systole nor diastole came to be, nor when nor where the dilatation and the constriction occurred, and that by reason of the quickness of the motion which in many creatures appeared and disappeared in the twinkling of an eye, like the passing of lightning, so that I thought that I sometimes saw systole on this side and diastole on that, sometimes quite the reverse, and sometimes the movements were changing and without order. And so I was much troubled in mind and did not know what to think, whether what I myself had concluded or whether to believe others, and I did not wonder at that which Andreas Laurentius had written, that the motion of the heart was like the ebbing and flowing of Euripus to Aristotle.[1]

At last, using daily more search and diligence, by often looking into many and different sorts of living creatures, by collecting and comparing many observations, I believed that I had hit the nail on the head, unwinded and freed myself from this labyrinth and had gained the knowledge I so much desired,

that of the movement and use of the heart and arteries. Since then I have not been afraid both privately to my friends and publicly in my anatomical lectures, after the fashion of the Academy,[2] to deliver my opinion on this matter.

As it commonly falls out, this opinion pleased some and displeased others. These stormed at me, slandered me and made a sin of it that I had departed from the precepts and belief of all anatomists. The others, avowing that it was a new thing, worthy of investigation and likely to prove most useful, asked me for a fuller explanation. At last, moved partly by the requests of my friends that they might all be made partakers of my endeavours,[3] and partly by the malice of some who, hearing with a biased mind what I had said and not understanding it, endeavoured to traduce me publicly, I was forced to commit these things publicly to print so that every man might pass his own judgement on me and on the business itself. But I do it the more willingly because Hieronymus Fabricius ab Aquapendente, having learnedly and accurately described in separate treatises almost all the parts of living creatures,[4] left only the heart untouched. Lastly, if anything of use or profit might accrue to the republic of letters from my endeavours in this matter, it might perhaps be granted that I had done well, and others might see that I had not spent my time altogether to no purpose, and as the old man says in the play:[5]

> No man so well e'er laid his plan to live,
> But that times, age, and use some new things give,
> And show you know not what you thought to know,
> And what you once thought best, you must forgo.

This may perhaps fall out now with regard to the motion of the heart, or at least, the way being thus laid down, others hereafter, trusting to more pregnant wits, may take occasion to conduct the business more advantageously and make better and further search.

NOTES

1 The whole of this paragraph shows the direct influence of the text of Andreas Laurentius, *Historia anatomica humani corporis*, Bk IX, cap. 10, and more particularly Bk IX, Q. 7, where Laurentius has written:

> But the nature and cause of its perpetual movement is entangled with so many layers of difficulties and such great ones, that the learned Fracastorius thought that it was known only to Nature and to God alone. I too think that the nature of this movement is no less worthy of admiration than that of the narrow strait of Euripus in Euboea which ebbs and flows at set intervals seven times by day and by night, the cause of which motion, Aristotle, while in exile in Chalcidia, could not discover, and so was stricken with grief and brought nigh to death.

Harvey used a copy of this work in the edition printed at Frankfurt in 1600. (He gives a page reference to it in his *Anatomical Lectures*.) It is not possible to say whether he went beyond it to the original writings of Fracastorius. Laurentius's reference to Euripus comes from Galen.

2 There is no particular reason to take this as a specific reference to the Academy founded by Plato about 385 B.C. and which became famous as a centre of Neo-platonism in the fifth century A.D. It seems rather more likely that Harvey is referring to the normal procedure of a university such as Padua where, having delivered a lecture, the professor was expected to 'dispute' his conclusions with the students and with his colleagues.

3 If one were to believe their authors, every anatomical textbook published in the sixteenth and seventeenth centuries was produced as the result of the persistent entreaties of pupils or friends.

4 It is not quite clear what Harvey had in mind when he wrote this. Fabricius had produced a number of monographs on various topics, all of which were designed to form part of a great study of the whole of the anatomy of the body, but he died before completing his project. It is true that he did not write on the heart, except on the foetal heart in his works on embryology, but he did not write either on the brain or on various other parts of the body.

5 Terence, *Adelphi*, V, 4, 1.

CHAPTER 2

Of the manner of movement of the heart as seen in the dissection of living creatures

First then, in the hearts of all creatures that are still living, after you have opened the chest and cut up the capsule which immediately surrounds the heart, you may observe that the heart moves sometimes, sometimes rests, and that there is a time when it moves and when it moves not.

All this is more evident in the hearts of colder creatures, as toads, snakes, frogs, snails, lobsters, crustaceans, molluscs, shrimps and all manner of little fish. Everything is also more evident in the hearts of warmer animals, like dogs and pigs, if you observe attentively until the heart begins to die and to beat more faintly and, as it were, to be deprived of life. Then you may clearly and plainly see the movements of the heart becoming slower and less frequent and its moments of stillness longer; and you may observe and distinguish more conveniently both the kind of movement that it has and how it is made. In its moment of stillness, as in death, the heart is limp, flabby, and listless, and lies as it were drooping.

In its movement and in the time of that movement, three things are chiefly to be observed:[1]

1. That the heart rises up and lifts itself upwards into a point, so that at that moment it strikes the chest and the beat can be felt outside.

2. That it is contracted on all sides, but more in width, so that it looks smaller in size and rather long and narrow. The

heart of an eel, taken out and laid on the table or on your hand shows this clearly; or again, it is equally apparent in the hearts of little fish and of those colder animals whose hearts are conical and rather elongated.

3. That the heart being grasped in one's hand while it is in motion feels harder. This hardness arises from tension, just as, if you grasp the muscles of the forearm with one hand while they are moving the fingers, you will feel them become more tense and more resisting.

4. It is, moreover, to be observed in fish and colder-blooded animals, like snakes and frogs and so forth, that at that time when the heart moves, it becomes paler in colour; when it lies quiet, it is to be seen richly-dyed blood-red in colour.

From these things it seemed evident to me that the movement of the heart was a kind of tension in every part of it, both along the lines of all its fibres as well as a contraction from all sides, because it seemed that in each motion it was lifted up, gained strength, decreased in size and hardened, and that this its movement was like that of the muscles when a contraction is being made along the length of the sinewy parts and of the fibres; for when muscles are moving and in action, they gain strength and become tense, from soft they become hard, they are lifted up and thickened, and so likewise the heart.[2]

From these observations we can say with good reason that the heart, when it is in motion, is everywhere contracted down and its walls are thickened, its ventricles are narrowed and it thrusts forth the blood contained within it. This is quite clear from the fourth observation, seeing that in its contraction it becomes paler in colour because it has driven out the blood it previously contained, and immediately after, in its relaxation and stillness, a purplish, blood-red colour returns to the heart as the blood enters once more into the ventricles. But of this no one need have any further doubts, for, if he make an incision into the cavity of the ventricle, he will see the blood contained in it spurt out violently with each movement or pulsation of the heart in its contraction.

So then, these things happen together at one and the same time: the tensing of the heart, the lifting up of its tip, the beat which is felt outwardly by reason of its hitting against the chest, the thickening of the walls of the heart and the forcible thrusting out of the contained blood by the constriction of the ventricles.

From this appears the reverse of the commonly received opinions, namely, the belief that at the moment when the heart strikes the chest and the pulse is felt outside, that at this same moment the ventricles are dilated and filled with blood, whereas you must realize that the contrary actually happens, and that when the heart is contracted, then it is emptied. And therefore, the movement which is commonly thought to be the diastole of the heart is really the systole. So likewise, the proper movement of the heart is not diastole but systole, and the heart does not increase in strength in diastole but in systole, for then it is tensed and moves and gains strength.

Nor is it in any wise to be allowed that the heart moves along the direction of its straight fibres only, although it has been supported by an analogy drawn by the great Vesalius with a wicker-work ring, meaning a bundle of rushes tied together in the shape of a pyramid.[3] According to this, while the apex of the heart approaches the base, the sides bulge out into a round and the cavities are dilated and the body of the heart acquires the shape of a little gourd and takes in the blood. But the heart is made tense and contracts along all the directions of all the fibres that it has at the very same moment, and its walls and its substance are thickened and enlarged rather than the ventricles. When the fibres running from the apex to the base contract, they immediately draw the tip towards the base;

the sides of the heart do not bulge out into a round, but rather the contrary, for each fibre set circular-wise when it contracts tends to become straight. And these circular fibres, like all the fibres of muscles, while they are contracted, they are shortened in length and so enlarged in width and are thickened after the same manner as the bellies of muscles.[4] To this add that in the movement of the heart, it happens that not only are the ventricles narrowed by reason of the straightening and thickening of their walls, but furthermore, because those fibres or fibrous bands which Aristotle called sinews[5] in which there are only straight fibres (in the walls all the fibres are circular), and which are diverse in the ventricles of the heart in larger animals, when they contract simultaneously, all the sides are forcibly drawn inwardly towards each other as if by a noose, in order to expel the blood with greater force, an arrangement worthy to be admired.

Neither is it true, as is commonly believed, that the heart by any motion or distention of its own draws the blood into the ventricles, for while it is moving and contracting then it expels the blood, and while it relaxes and subsides it receives the blood in a way which will presently be made plain.

NOTES

1 This chapter repeats in greater detail observations that Harvey had already made by 1616 and which he records in his *Anatomical Lectures.* The four points which he makes here will all be found in the *Lectures*, though not arranged with equal cogency (see *Anat. Lect.*, pp. xlii–xliii). The first anatomist to give a correct description of the systole and diastole of the heart was Realdus Columbus in his *De re anatomica*, Bk XIV, p. 257, and Harvey quotes the passage at length in his *Lectures*, p. 265. Apart from Volcher Coiter who tried hard to watch the movements of the heart but ended in a muddle, no anatomist, as far as I know, between Columbus and Harvey realized that Columbus was right, and that systole and diastole were the reverse of what was commonly supposed. (See *WH and the Circulation*, pp. 70–6.)

2 In *De motu locali animalium*, Harvey had considered at length the problem of the contraction of muscle and had reached the conclusion that as a muscle contracts, it shortens in length and increases in width and breadth in the belly, that it contracts towards the belly and that contraction is most powerful in the belly of the muscle (see especially pp. 121–3).

3 Vesalius's description of the manner of the heart's action (Bk VI, cap. 15) is based upon his belief in the attractive power of the heart, that is, its power to attract blood from the vena cava into the right ventricle and air from the pulmonary vein into the left, at the moment when the heart is in active dilatation. This attractive power he describes by various analogies: smiths' bellows sucking in air, or 'after the manner of the flame of lanterns attracting oil', or again, 'as the magnet attracts iron by the intimate similarity of its quality, for what is more intimately similar to the heart itself than the air destined for its refrigeration and the nurturing of its innate heat?' (p. 598). And he thinks of the air as rushing in to prevent a vacuum from occurring in the distending heart. He begins his description of the heart's action by saying that the substance of the heart, which is similar to that of muscle, is peculiarly designed for the performance of its special functions, that is the 'alteration' (in the Aristotelian sense) of blood for the sake of the lungs and the preparation of vital spirit. The fibres of the heart serve the alternate movements of dilatation and contraction and the period of quiescence between them. But whereas the movement of muscles is voluntary, that of the heart, being incessant, is natural and outside the control of the nerves. He then describes the dilatation of the heart as the drawing up of the tip towards the centre of the base and the distending of the sides of the heart, and says that it is effected by the straight fibres which contract and pull the tip towards the base. This action he believes corresponds to the enlarging of the internal cavities of the heart, and this he tries to prove by his mechanical model. If a bunch of rushes be fixed by their stems around a circular osier and their tips tied together so that they form a kind of pyramid, and if then the tip be pushed towards the base, the pyramid will become shorter, but the space inside will become larger. The contraction of the heart he explains as the result of the relaxation of the straight fibres and the contraction of the transverse or circular fibres, and again uses the example of the bunch of rushes, saying that if you encircle it about the middle with the hands and squeeze it, the tip of the pyramid will recede from the base, the internal cavity will become smaller and the whole bunch longer in length. For the action of the

oblique fibres he can give no model, but merely says that they act during the quiescence of the heart. He concludes that the straight fibres serve for attraction, the circular for expulsion and the oblique for retention.

This description differs only in detail from that given by Vesalius's predecessors. Clearly Vesalius had no idea of the true action of the heart or of the correct timing of its systole and diastole, and one may well ask what he really thought about the timing of the systole and diastole of the arteries.

4 *Anat. Lect.*, p. 271: 'when the fibres of the heart are contracted lengthways the walls of the heart are thickened, when the fibres are contracted widthways the walls are compressed as happens in the muscles of the abdomen.'

5 *P.A.* 666b 12: 'The heart is abundantly supplied with sinews.' Under the word *neura*, Aristotle included tendons, sinews, ligaments and all other fibrous parts, and also any nerves that he may have noticed. The active part in muscular contraction he held to be provided by the *neura*, the tendinous fibres. On the action of these 'fibrous bands', see below, Chapter 17, pp. 123–4.

CHAPTER 3

Of the manner of the movement of the arteries as seen in the dissection of living creatures

In the movement of the heart there occur these further things which must be observed and which relate to the movements and pulsation of the arteries.[1]

1. At the moment when the heart is tense, makes a contraction and strikes the chest and is altogether in systole, the arteries are dilated, give forth a pulse and are in their diastole. In like manner, when the right ventricle contracts and thrusts out the blood contained in it, the pulmonary artery pulsates and is dilated simultaneously with the rest of the arteries of the body.

2. When the left ventricle ceases to move, beat and be contracted, the beating of the arteries ceases; indeed, when it contracts more feebly, the pulse in the arteries is hardly to be perceived. So likewise it happens in the pulmonary artery when the right ventricle ceases to pulsate.

3. Again, if any artery be cut or pierced, the blood is forcibly thrust out of the wound at the moment of contraction of the left ventricle. So also, if the pulmonary artery be cut, you will see the blood burst out forcibly from it at the very moment when the right ventricle becomes tense and contracts.

So likewise in fish, if you cut the vessel which leads from the heart to the gills, you will also see the blood forcibly thrust out through the cut at the very moment that the heart is tensed and contracted.

And lastly, since in any arteriotomy the blood leaps out

sometimes to a greater distance and sometimes to a less, you can find out for certain that this leaping occurs in the diastole of the arteries, at the time when the heart strikes the chest; and this undoubtedly happens at the time when the heart is seen to be tensed and contracted and to be lifted up in its systole, and at the same time and by this same movement the blood is driven forth.

From these things it is clear that, contrary to what is commonly believed, the diastole of the arteries occurs at the same time as the systole of the heart, and that the arteries are filled and distended by reason of the inflowing and inthrusting of blood made by the constriction of the ventricles; as likewise, that the arteries are distended because they are filled like water-skins or bladders, and they are not filled because they are distended like a pair of bellows. And all the arteries of the whole body pulsate for the same cause, that is, by reason of the contraction of the left ventricle of the heart, just as the pulmonary artery pulsates by reason of the contraction of the right ventricle.

Therefore, the pulsation of the arteries arises from the impulsion of blood from the left ventricle. It is after the same fashion as when one blows into a glove, all the fingers are distended at the same time and mimic the impulsion of the air.[2] And indeed, in accordance with the tension of the heart, so the pulsations are in an equal degree larger, more vehement, frequent or swift, keeping like rhythm, quantity and order with the heart. Nor is it to be expected that on account of the movement of the blood there should be a certain lapse of time between the constriction of the heart and the dilatation of the arteries, especially of those that are the furthest distant, or that they do not occur at the same instant, for the case here is the same as in blowing up of a glove or bladder, or striking a plenum like a drum or a long piece of timber when the blow and the resulting motion are felt simultaneously at both extremities.[3] And as Aristotle says: 'The blood of all living creatures beats within their veins (meaning their arteries), and by the pulsation is moved everywhere at the same time.'[4] And, 'so do all the veins beat together and in their turn, because they all derive from the heart. The heart always moves, wherefore they likewise always move in their turn and all at the same time whenever the heart moves.'[5]

We must observe with Galen that the arteries were called veins by the ancient Philosophers.

I chanced once to see and have at hand, a particular case which confirmed the truth of this for me most clearly. A certain person had a great pulsating tumour, called an aneurysm, on the right side of his neck, near the place where the subclavian artery descends to the axilla. It was derived from an erosion of the artery itself and grew bigger and bigger every day, being filled with the immission of blood from the artery at every pulsation, as we found when his body was dissected after death. In this man, the pulse in his arm on that same side was very weak, because the greater part of the inflowing blood was diverted into the tumour and so intercepted.

Wherefore, whether it be by compression or infarction or interception, in any place whatsoever, that the movement of the blood through the arteries be hindered, then the arteries beyond this place pulsate less, seeing that the pulse of the arteries is nothing but the impulsion of blood into the arteries.

NOTES

1 Most of the observations recorded in this chapter also had been made by Harvey before 1616 and will be found in the *Anatomical Lectures* (see p. xliii). Galen had taught that the systole and diastole of the heart was synchronous with the systole and diastole of the arteries. As few people, if any, apart from Columbus, had a correct understanding of the timing of the

systole and diastole of the heart, they were all equally in the dark about the systole and diastole of the arteries. Vesalius was well aware that the arterial pulse was occasioned by the action of the heart but he does not seem to have made any clear statement on the timing of the heartbeat and the arterial pulse. His readers are merely bidden to observe these things for themselves in a living dissection. (See *First Public Anatomy*, 1540, p. 293; *Fabrica*, p. 662, misnumbered 658; p. 660.)

2 *Anat. Lect.*, p. 273: 'Action of the heart: thus relaxed receives blood, contracted scups it over. The whole of the body responds to the artery as my breath in a glove.'

3 This statement is not correct. The pulse felt peripherally, for example in the foot or at the wrist, is not synchronous with the pulse of the heart felt through the chest-wall. The ejection of blood into the aorta produces a wave of increased pressure, the pulse wave, which travels through the arteries to the periphery at a velocity of approximately 5–6 m per sec. This pulse wave is simply a pressure wave and is not to be confused with the velocity of the blood flow through the arteries which nowhere exceeds 0.5 m per sec. at rest and is considerably less in the smaller vessels.

4 *H.A.*, III, 19, 521a 6: 'Blood beats or palpitates in the veins of all animals alike all over their bodies. . . .'

5 *De respirations*, cap. 26 (20), 480a 10: 'All the veins pulse, and do so simultaneously with each other owing to their connexion with the heart. The heart always beats, and hence they also beat continuously and simultaneously with each other and with it.'

CHAPTER 4

Of the manner of the movement of the auricles of the heart as seen in the dissection of living creatures

Besides these things about the movement of the heart, there are others to be observed which touch the use of the auricles.

Caspar Bauhin and Jean Riolan, both very learned men and most skilful anatomists, have observed and inform us that if, in the live dissection of any animal, you pay careful attention to the movement of the heart, you will see four motions distinct in time and place, of which two are proper to the auricles and two to the ventricles.[1] Now, with the leave of such excellent men be it spoken, there are four motions distinct in place but not in time. For both the auricles move together and both the ventricles, so that there are four motions distinct in place, but only at two times. And it happens in this way.

There are as it were at one time two motions, one of the auricles and the other of the ventricles, and they are not altogether simultaneous. But the motion of the auricles goes before and the motion of the ventricles follows. And the motion seems to begin from the auricles and to pass forward to the ventricles. When all things are already in a languishing condition, the heart dying away, both in fish and in colder-blooded animals, there intercedes between these two motions a short time of stillness, and the ventricle being as it were awakened seems to answer to the motion sometimes swifter, sometimes slower, and at last, drawing towards death, it ceases to answer by its motion and only by gently nodding its head seems as it were to give consent and, moving scarce perceptibly, seems only to give a sign of motion to the beating auricle. So the ventricles cease to beat before the auricles, so that the auricles are said to outlive them. The left ventricle ceases to beat first of all, then its auricle, then the right ventricle, last of all (which Galen[2] also observed), all the rest now still and dying, the right auricle beats still, so that life seems to remain last of all in the right auricle. And whilst by little and little the heart is dying, you may see after two or three beatings of the auricles, the heart will, being as it were roused, answer, and very slowly and with difficulty bestir itself and beat once.

This is especially to be observed, that after the ventricle has ceased to beat and the auricles are beating still, if you put your finger on the ventricle you will perceive each pulsation of the auricle in clearly the same way as I said before that the pulsations of the ventricles are felt in the arteries, from the distension caused without doubt by the impulsion of blood. And at this time, while the auricle alone is beating still, if you cut away the point of the heart with a pair of scissors, you will see the blood flow from thence at every pulsation of the auricle, so that from thence it appears which way the blood comes into the ventricles, not by attraction or distension of the ventricles, but sent in by the impulsion of the auricles.

It is to be noticed that what I call pulsations, both in the auricles and in the ventricles, are contractions. And you will clearly see that the auricles are contracted first, and afterwards the ventricles. For while the auricles move and beat, they become whiter, especially when there is little blood in them. (They are indeed filled like a reservoir or lake of blood by the spontaneous tending of the blood to its centre and squeezed by the movement of the veins.)[3] Furthermore, it is most evident from the ends and extremities of the auricles that this whiteness arises from their contraction.

In fish and frogs and the like, creatures having but one ventricle of the heart and in lieu of an auricle a little bladder crammed full of blood placed at the bottom of the heart, you will most plainly see this bladder contract first, and afterwards be followed by the contraction of the ventricle.

However, it seems to me that I should here insert those things which I have observed and which happen in a contrary manner. The heart of an eel and of certain other fish and animals, being taken out of the body, beats without auricles. Furthermore, if you cut it in pieces, you will see the separate pieces each contract and relax, so that in them the very body of the heart beats and leaps after the auricles have ceased to move. But whether this is peculiar to creatures that are more tenacious of life, whose radical moisture is more glutinous, or fat and sticky and not so easy to be dissolved, is hard to tell. This same thing appears also in the flesh of eels, for after they have been skinned and degutted and cut into pieces, their flesh still retains a movement.

It is certain that once, when I was experimenting on a pigeon, after the ventricles had quite ceased to move and even the auricles were already flagging, I wetted my finger with spittle and kept it for a while, warm and wet upon the heart. As if it had recovered its strength from this fomentation, and was resuming its former life, I saw the heart and its auricles begin to move, to contract and relax, and it seemed, as it were, called back again from the realms of the dead.

Besides all these things, I have observed at different times that after the heart itself and even its right auricle had left off beating at the very point of death, there manifestly remained in the very blood itself which is contained in the right auricle, an obscure motion and a kind of flooding and beating for as long, that is to say, as it seemed to be imbued with any warmth and spirit.[4]

A thing of like nature is very clearly to be seen in a hen's egg, in the first foundation of the chick within seven days of incubation. First of all there is in it a drop of blood which beats, as Aristotle likewise observed,[5] from which, when it has grown further and the chick has been formed in part, the auricles of the heart are fashioned, and in their incessant beating life is present. Within a few days afterwards, the outline of the body begins to take shape and the heart itself is created and for some time appears white and bloodless like the rest of the body, and it neither beats nor moves. Indeed, I have also seen in a human foetus at about the beginning of the third month, the heart formed, but white and bloodless, but in its auricles, notwithstanding, great store of blood, crimson in colour. So likewise in an egg, when the chick was formed and grown, the heart also began to grow and to have ventricles into which it received the blood and then sent it forth.

So that if any man will look more deeply into the truth, he will not say merely that the heart is the first thing that lives and the last that dies, but its auricles (and in snakes, fish and suchlike creatures, the part which serves as an auricle), and that its auricles have life before the ventricles and die after them.

Furthermore, there is some doubt as to whether or not, before this, the blood or spirit has in itself the obscure beating which to me it seemed to retain after death, and whether we should say that from this beating life begins, seeing that the sperm of all living creatures and their prolific spirit goes forth from them with a kind of leaping as if itself were a living creature, as Aristotle observed.[6] As if Nature should thus act in reverse, as Aristotle says,[7] and having made a descent into death, betake herself with backward steps from the finish to the starting pits from which she rushed forth, and since the generation of every animal proceeds from what is not animal into an animal, as if from a non-being into a being, corruption passes from a being into a non-being by the same steps in reverse; whence it is that that which in living creatures is last made fails first, and that which is first made fails last.

I have likewise observed that there is indeed a heart in almost all animals and not only, as Aristotle says,[8] in those of the larger sort and that are sanguineous, but also in the smaller bloodless ones, crustacea and certain testacea, such as slugs, snails, molluscs, crayfish, lobsters, shrimps and many others; indeed, in wasps, hornets and flies even, with the aid of a magnifying glass made to reveal the smallest things, I have seen the heart beating at the top of that part which we call the tail, and I have shown it to others.[9]

But in bloodless creatures the heart beats very slowly and with infrequent strokes, as it does in other creatures that are dying, and it contracts leisurely as can easily be seen in snails. You will find their heart on the right side at the bottom of the orifice which seems to open and shut for the sake of ventilation, and from whence the slime is cast out, if you open the body at the top, near the part which corresponds to the liver.[10]

Now this also should be noticed, that in winter and colder seasons, some bloodless creatures, such as is the snail, have no pulsation but seem rather to live the life of plants, like those other creatures which for this very reason are called plant-animals.[11]

Furthermore, it is to be observed that all creatures which have a heart, have likewise auricles or something which corresponds to auricles; and that wheresoever the heart has two ventricles, there are always two auricles, but that the reverse is not so. If you look closely at the formation of the chick in the egg, you will see, as I have said, that first of all there is in it only a bladder or auricle or drop of blood which beats, and later, when it has grown, the heart is finished. So in some creatures which do not, as it were, acquire any further perfection, like bees, wasps, snails, shrimps, lobsters and so on, there is only a pulsating vesicle like a tiny point, flashing now red, now white, as the beginning of life.

There is here with us a sort of very little fish called a shrimp (in Low Dutch *een Garneel*), which is usually taken in the sea and in the Thames, and all its body is transparent. This little fish I have often shown in water to some of my special friends so that we could most clearly discern the motion of the heart in the creature, for the outward parts of its body did not obstruct our sight at all, and we could see the beating of the heart as if it had been through a window.[12]

In a hen's egg, some four or five days after incubation, I showed the first beginning of the chick like a little cloud, by putting the egg off which the shell was taken, into water warm and clear. In the midst of the cloud, there was a point of blood which did beat, so little that when it was contracted, it disappeared and vanished out of our sight, and in its dilatation, showed itself again red and small as the point of a needle, so that between being seen and not being seen, as it were between being and not being, it represented a beating and the beginning of life.

NOTES

1 Bauhin, *Theatrum anatomicum*, II, 21; Riolan, *Anthropographia*, VIII, 1. Harvey is quoting Bauhin here verbatim. This analysis of the heart's movement will not be found in the *Anatomical Lectures*. For the points in common between them and this chapter, see *Anat. Lect.*, p. xliv.

2 *De anatomicis administrationibus*, VII, 15; '. . . in each ventricle the apex stops moving first and then the part next to it, and so on until the bases only are left still moving. When even these have stopped, an ill-defined and short

movement at long intervals is still seen in the "auricles".' See Singer's trans., *On Anatomical Procedures*, p. 196.

Harvey's account of the dying heart is perhaps one of the most brilliant pieces of observation in the whole book and the most graphically written. As Dr K. D. Keele pointed out (*William Harvey*, p. 129), he describes all the stages of partial and complete heart-block from the increased interval between the contractions of the atria and the ventricles to their final complete dissociation.

3 See below, Chapter 15, p. 111.

4 Harvey had presumably seen the effects of the terminal fibrillation of the atrial wall.

5 *H.A.*, VI, 3, 561a 12. Harvey's detailed observations on the early stages of the developing chick embryo will be found in his *De generatione*, Ex. XVII. See also below, Chapter 16.

6 *De motu animalium*, 703b 25. See also Harvey's *De motu locali animalium*, p. 35: 'The first instrument of movement is spirit. And the natural causes of movement are heat and cold; hence the semen goes forth as if itself a living creature, and hence too the movement of the heart.' Harvey returns to the question of the 'obscure' palpitation in the blood on several occasions, notably in his Second Letter to Riolan and in his *De generatione*, in particular in Ex. LI. See also *WH and the Circulation*, pp. 232–4.

7 *D.G.A.*, II, 6, 741b 20–5. The analogy which Harvey uses derives from the chariot races which set out from the *carceres* or 'starting pits'. It is used by Cicero, *De senectute*, with the meaning of 'to start life anew'.

8 *P.A.*, III, 4, 665b 10.

9 Harvey's observations on insects were among the papers pillaged from his lodgings in Whitehall by the Parliamentary soldiery, with the result that it is only from remarks such as these scattered throughout his works that we can now have any idea of their extent and scope. It is generally supposed that Marcello Malpighi was the first to idenitfy the heart of an insect. See E. J. Cole, *A History of Comparative Anatomy*, London 1944, p. 187.

10 Some of the earliest work on the anatomy of the snail was done by Jan Swammerdam 1675, Francesco Redi 1681 and Martin Lister 1692. The latter describes the heart as having a cleft or bifid single ventricle and says that Harvey is 'false and assuredly contemptible' in thinking that the heart of the snail consists in an auricle without a ventricle. See Cole, *op. cit.*, pp. 233–4, 237.

11 For Harvey's further remarks on the subject, see below, Chapter 17, p. 120.

12 In his *Anatomical Lectures*, Harvey had observed that the heart of a shrimp is white (pp. 263, 273), but made no reference to its movement.

CHAPTER 5

Of the action and function of the movement of the heart

I confidently believe then that at last, out of these and the like observations it will be found that the movement of the heart is after this manner.

First of all the auricle contracts itself, and in that contraction throws the blood with which it abounds as the head of the veins and the storehouse and cistern of the blood, into the ventricle of the heart, which being filled, straightway the heart raises itself, tenses all its fibres, contracts the ventricles and makes a beat, by which beat it forthwith thrusts the blood sent in from the auricle out into the arteries: the right ventricle into the lungs through the vessel called the arterial vein [pulmonary artery] but which is indeed by its constitution, its function and by all else an artery; the left ventricle into the aorta and so through the arteries into the whole body.

Those two motions, the one of the auricles, the other of the ventricles, follow so hard upon each other's heels, keeping as it were, in harmony and rhythm,[1] that they occur simultaneously and only one motion is apparent, especially in warmer animals where they are driven with a rapid motion. Nor is this otherwise done than when, in machines, one wheel moves another and they all seem to move together;[2] or in that mechanical contrivance which is fitted to firearms where, by compressing the trigger, the flint falls, strikes forcibly upon the steel and brings forth a spark which falls onto the powder which is ignited, enters the touch-hole and explodes, and the bullet flies out and pierces the mark, and all these movements by reason of their swiftness appear to happen simultaneously as in the twinkling of an eye. So likewise in swallowing, the food or drink is thrown into the gullet by the elevation of the root of the tongue and the compression of the mouth, the larynx is closed by its own muscles and by the epiglottis, the top of the gullet is lifted up and opened by its muscles, just as a sack, is raised to be filled and its mouth opened wide to receive the filling, and the food or drink being received, it is pressed downwards by the transverse muscles and pulled down by the longer ones. And yet, notwithstanding that all these motions are made by several and contradistinct organs, whilst they are done in harmony and order, they are seen to make but one motion and action which we call swallowing.

It clearly happens thus in the motion and action of the heart, which is a kind of swallowing and transfusion of the blood from the veins into the arteries. And if any man, bearing these things in mind, will diligently observe the motion of the heart in a live dissection, he will see not only what I have said, that the body of the heart lifts itself up and makes one movement which is continuous with that of the auricles, but also a certain undulation and obscure lateral inclination along the line of the right ventricle, as if the heart twisted itself slightly to accomplish the business.[3] And just as we may see when a horse drinks and swallows water, at every gulp the water is supped down into the belly and this movement makes a noise and yields a pulse to those who listen and touch him, so also when some portion of the blood is transferred from the veins to the arteries by these movements of the heart, a pulse is made and can be heard within the chest.[4]

The motion of the heart then is after this manner and one of the actions of the heart is the very transmission of the blood and its propulsion to the extremities by the intermediacy of the arteries, so that the pulse which we feel in the arteries is nothing else but only the impulsion of blood by the heart.

Now whether the heart besides transferring the blood, giving it local movement and distributing it, adds anything further to it, as heat, or spirit, or perfection, we must enquire hereafter and consider from other observations.[5] Let it suffice for the present that it is sufficiently shown that in the beating of the heart, the blood is transferred and brought out of the veins into the arteries through the ventricles of the heart and so distributed into the whole body.

But this indeed all men do in some manner concede and conclude from the fabric of the heart and from the skilful contrivance, siting and use of its valves. Yet, stumbling as it were in a dark place, they seem to be dim-sighted and clumper up diverse things which are contrary and inconsistent and speak many things by guess as I have shown before.[6]

One thing seems to me to have been the chief cause of doubt and mistake in this business, and that is the connection in man of the heart and the lungs. For when they had seen the pulmonary artery and the pulmonary vein likewise disappear into the lungs, they were much at a loss to see how the right ventricle should distribute the blood into the body, or how the left ventricle should draw it out of the vena cava. Galen's words bear witness to this in *De placitis Hippocratis et Platonis*, Bk VI, where he inveighs against Erasistratus concerning the origin and use of the veins and the concoction of the blood, saying:

> You will answer that it is so ordained, that the blood be prepared in the liver and from thence carried to the heart, there to receive its proper form and absolute perfection. Now this does not seem to lack reason, for no perfect and great work is done suddenly, at one attempt, nor can it acquire all its refining from one instrument. If it indeed be so, show us another vessel which draws out the blood,

being absolutely perfected, from the heart and dispenses it into the whole body as the artery does the spirit.

Here you can see a reasonable opinion left and rejected by Galen, because, besides not seeing the passage for the transference of the blood, he could not find a vessel which should distribute the blood from the heart into the whole body.

But if anyone had spoken at that time in defence of Erasistratus, or of that opinion which we now hold, and which is in other ways conformable to reason, as Galen himself confesses, and had pointed out with his finger the great artery dispensing the blood from the heart into the whole body, I wonder what that great man, so clever and so learned, would have replied? If he had said that the artery distributed spirit and not blood, he would indeed have refuted Erasistratus sufficiently well, for Erasistratus believed that spirit was merely contained in the arteries; but in the meantime, he would have contradicted himself and basely denied the very thing which in his own book he forcefully maintained against this same Erasistratus, and proved with many valid arguments and showed by his experiments, that it is blood which is naturally contained in the arteries and not spirit.

But if great Galen would admit, as he does often in this same book, that 'all the arteries of the body arise from the great artery and it from the heart', and further that 'in all these arteries blood is naturally contained and carried', and would acknowledge that 'the three sigmoid valves placed in the orifice of the aorta prevent the return of blood into the heart', and that Nature would in no wise have assigned them to the most noble of our entrails unless they were to be employed for some most important office, if, I say, the Father of the Physicians would grant all these things and in the very same words as he does in the book I have quoted, I do not see how he could possibly deny that the great artery is a vessel of this kind that distributes the blood, after it has received its absolute perfection, out of the heart into the whole body. Or would he perhaps still hesitate, as all men have done since his time to this very day, because, on account of the connection of which I spoke between the heart and the lungs, he could not see the ways by which the blood could be transferred from the veins into the arteries.

This uncertainty does not a little perplex the anatomists when in dissections they always find the pulmonary vein and the left ventricle of the heart full of thick, knotty, black blood, so that they are forced to affirm that this blood sweated through the septum of the heart from the right to the left ventricle. But this way I have already refuted. Therefore another way must be prepared and laid open, which being discovered, I believe no further difficulty will then arise to prevent any man from granting and readily acknowledging those things which I have propounded before of the beating of the heart and arteries, of the transference of the blood from out of the veins into the arteries and of its distribution through the arteries into the whole body.

NOTES

1 For Harvey's views on harmony and rhythm in the body, see *De motu locali animalium*, pp. 67, 111, 143–9, 153.

2 Cf. Aristotle, *De mundo*, 398b 15–30; *De motu animalium*, 701b 1–5.

3 *Anat. Lect.*, p. 267: 'when the heart replies tardily to the movement of the auricle it twists its point a little towards the left'.

4 Probably the first reference ever made to the existence of cardiac sound, that is chiefly the noise made by the closing of the valves of the heart.

The first man to realize its significance and use in diagnosis was Laënnec in his *Traité de l'auscultation médiate*, published in 1819.

5 See below, Chapter 8, p. 76. For a discussion of Harvey's views on innate heat, see *WH and the Circulation*, pp. 219–35.

6 In the Introductory Discourse.

CHAPTER 6

Of the ways by which the blood is carried from the vena cava into the arteries, or out of the right ventricle of the heart into the left

Since it is probable that the connection which they see in man of the heart with the lungs has been the cause, as I have said, of their error, they are to blame in this, that whilst they desire to pronounce upon and demonstrate and understand the parts of all living creatures, as all anatomists commonly do, yet they look only into man's body, and that his dead body, and so do no more to the purpose than those who seeing the manner of government in one Commonwealth draw up a whole political system, or those who knowing the nature of one field believe that they understand agriculture; they act no otherwise than as if from one particular premiss, they would frame a syllogism to a universal conclusion.[1]

Were they, however, as well practised in the dissection of animals as they are experienced in the anatomy of man's carcase, this matter which keeps them all in doubt and perplexity would, in my opinion, shine forth clear and free of all difficulty.[2]

First of all then, in fish which have but one ventricle of the heart, they having no lungs, the matter is clear enough. For it is certain that the bladder of blood set at the base of the heart and analogous to an auricle sends the blood into the heart, and that the heart then clearly sends on the blood again through the pipe or artery or vessel analogous to an artery; and this can be confirmed before our eyes either by looking or by the cutting of that artery from which the blood then leaps forth at every pulsation of the heart.[3]

Next, it is not difficult to see the same thing in all animals that have but one ventricle, or as it were but one, as toads, frogs, snakes and lizards, which, although they are said in some manner to have lungs because they have a voice (about the wonderful contrivance of their lungs and about other things of this kind I have very many observations by me,[4] but this is not the place for them), yet it is plainly to be seen from actual inspection that in them the blood is transferred in the same way from the veins into the arteries by the pulsation of the heart and the way of it open, clear and manifest, presenting no difficulty and no occasion for doubt. For in these animals the case is as it might be in man were the septum of his heart perforated or taken away or one ventricle made out of the two; that done, I believe no man would then doubt by which way the blood could pass out of the veins into the arteries.

And seeing there are more creatures which have no lungs than there are which have them, and more which have but one ventricle of the heart than there are which have two, it is easy to conclude that in animals for the most part, ἐπὶ τὸ πολὺ, usually and generally, the blood is sent from the veins into the arteries by an open way through the cavity of the heart.

I have, moreover, considered in my own mind that the same thing is most clearly to be seen in the embryos of those animals that have lungs. In the foetus, there are four vessels of the heart, the vena cava, the pulmonary artery, the pulmonary vein and the aorta or great artery, and they are connected otherwise than in one come to age, as all anatomists know well enough.

The first contact and union of the vena cava with the pulmonary vein, which happens before the vena cava opens into the right ventricle of the heart or sends out the coronary vein, a little above its going out from the liver, employs a lateral anastomosis, that is a hole, wide open and large, of an oval shape, bored through from the vena cava into the pulmonary vein and affording a passage way, so that through this hole, as through a single vessel, the blood can most freely and abundantly pass out of the vena cava into the pulmonary vein[5] and into the left auricle of the heart and so into the left ventricle. There is, moreover, in that oval foramen, on the side facing the pulmonary vein, a fine hard membrane, larger than the foramen and like a lid,[6] which later, in the fully grown, covers up this hole and unites all round with its edges and entirely blocks and almost obliterates it. This membrane then is so constituted that, hanging loosely with its own weight, it is easily bent back and yields a passage to the blood flowing in from the vena cava to the pulmonary vein and to the heart,[7] but hinders it from flowing back into the vena cava again. So that from this we have reason to think that in the embryo the blood must incessantly pass through this foramen from the vena cava into the pulmonary vein and from thence into the left auricle of the heart, and being entered there can never return.

The other union is that of the pulmonary artery, and it happens after that vessel having issued from the right ventricle is divided into two branches, and it is, as it were, a third branch added to these two, and like an arterial canal, and it runs obliquely from here into the aorta which it perforates, so that in the dissection of embryos there appears to be, as it were, two aortas, or two roots of the great artery arising from the heart. This canal likewise is attenuated little by little in those come to age, and dwindles away, and at last is quite dried up and disappears like the umbilical vein. This arterial canal has no membrane in it to hinder the backward or forward movement of the blood. For there are in the orifice of that pulmonary artery of which, as I have just said, this canal is a branch, three sigmoid valves which open from within outwards and easily give passage to the blood flowing by this way from the right ventricle into the great artery, but on the contrary completely prevent anything from returning either from the aorta or from the lungs into the right ventricle which is perfectly closed by them. And so here again we have reason to think that in an embryo, when the heart contracts, the blood is incessantly carried by this way from the right ventricle into the great artery.

What is commonly said, that these two unions, so great, so wide open and so clear, were made only for the nourishing of the lungs and that they are abolished and closed up in those come to age, that is at a time when the lungs require a more abundant nutriment on account of their own heat and movement, is a fiction both improbable and inconsistent. And this is likewise false which they say of the heart in an embryo, that it keeps holiday, does nothing and moves not at all,[8] and therefore Nature was forced to make those passages for the nourishing of the lungs. For it is plain enough if we look with our own eyes into an egg on which a hen has sat, and into embryos newly brought forth from the womb, that the heart moves as it does in the fully grown, and so Nature was compelled by no such necessity. Of this movement, not only these eyes of mine have been the frequent witnesses, but Aristotle himself attests it, saying: 'The pulse appears immediately from the very beginning in the fashioning of the heart and this is to be discovered in the dissection of living creatures and in the formation of the chick from the egg.'[9] In truth, I have seen that these passages are open and free, both in the race of men and in other animals, not only up to the time of birth, as the anatomists have remarked, but also for many months after, in some indeed for many years if not for all their lifetime, as in the goose, woodcock and very many birds and animals, especially the smaller ones.[10] And this, perhaps, deceived Botallus[11] who thought that he had found a new passage for the blood from the vena cava into the left ventricle of the heart, and I confess that when I myself first found it in a rather large, full-grown mouse, I at once thought it to be something of this kind.

From these things it is to be understood that, in the human embryo and in other animals likewise in which these conjoinings are not obliterated, this same thing happens, that the heart by its own motion transfers the blood most clearly from the vena cava into the great artery by very patent ways through the ducts of both the ventricles. For the right ventricle, receiving the blood from its auricle, thrusts it forth through the pulmonary artery and its branch, called the ductus arteriosus into the great artery. The left ventricle likewise, at the same time, as a result of the movement of its auricle, receives the blood which has been brought into that left auricle from the vena cava through the foramen ovale, and by its tensing and contraction immediately drives it on in the same way through the root of the aorta into the great artery itself.

And so in embryos, while the lungs are idle and have no action nor motion (as if they did not exist), Nature makes use of both the ventricles of the heart as if they were but one for the transmission of blood. And so the condition of embryos that have lungs but as yet make no use of them, is the same as that of other animals that have no lungs at all.

Therefore, in embryos also, the truth shines forth as clearly, the truth that the heart by its beating transfers the blood out of the vena cava into the great artery and pours it through as patent and open ways as if in man, as I said before, both the ventricles were made pervious to each other by taking away the septum between them. Seeing therefore, that in most animals, and in all for a certain time, there exist these most open ways that serve for the transmission of blood through the heart, it remains for us to seek diligently into this business. Either why in some animals, as in man and in those that are warmer-blooded and come to full growth, we do not believe that the same thing is performed through the substance of the lungs which Nature accomplished before in the embryo through those passages at the time when the lungs had no use, passages which she seems to have been forced to make for want of a way of passing through the lungs. Or why it is more profitable, for Nature always does that which is more profitable, that Nature should have completely closed to the passage of the blood in those that are growing up, those open ways of which she made use before in the embryo and in the foetus, and in all other creatures continues to use, without opening any other ways for that passing across of the blood, but instead altogether preventing it in this manner.

So then, the business is now come to this, that for those who seek for the ways in man by which the blood flows through from the vena cava into the pulmonary vein and into the left ventricle, it were more worthy their pains and would seem more advantageous, if, from the dissection of animals, they were willing to search out the truth and enquire the reason why, in the larger and more perfect animals and in those that are fully grown, Nature preferred to have the blood strained through the parenchyma of the lungs rather than through the most open passages, as in other creatures, and then they would understand that no other way nor passage could possibly be thought of. They should enquire whether this be because the larger and more perfect animals are hotter, and when they are full grown their heat is more fiery, so to speak, and more apt to be suffocated, and therefore the blood permeates the lungs and is thrust through them, that its heat may be tempered by the air that is breathed in and the blood kept from boiling and the heat from suffocation, or whether it be for some other reason of the same kind. But to determine these things and give a complete answer is nothing else but to seek out the cause for which the lungs were made. And of the lungs, their use and movement, and of all manner of cooling, of the necessity and use of air and of other things of this kind and of the various organs and the differences made in them in animals on this account, although I have discovered many things from the countless observations I have made, yet I shall leave them to be more conveniently set forth in a treatise by themselves,[12] lest by straying here too widely from my subject of the movement and use of the heart, I should be thought to go aside from my intention and leave my stance and confuse and evade the problem. And so, that I may return to my proposed goal, I will go on to prove what remains.

And first I prove that in the more perfect and warmer-blooded animals and in those come to age, as in man, the blood passes from the right ventricle of the heart through the pulmonary artery into the lungs, and from thence through the pulmonary vein into the left auricle and then into the left ventricle of the heart. First I will show that this may be so, and then that it is indeed so.

NOTES

1 Cf. *De generatione animalium*: 'fond and erroneous is that method of seeking truth in use in our times, according to which most men diligently enquire not what a thing is, but what other men say it is, and inferring a universal conclusion from particular premisses and from thence often shaping for themselves a false analogy, they for the most part pass on to us things like truth for truth itself' (Praefatio, C).

2 From at least the time of Vesalius onwards, it was common practice in Padua to dissect animals to supplement human material and to demonstrate many physiological problems from vivisected dogs. This practice, however,

does not seem to have been followed in Oxford or Cambridge or in the Colleges of Physicians and Surgeons of London. The first anatomists to make any serious study of animal anatomy as such were Fabricius ab Aquapendente, his pupil and successor at Padua, Julius Casserius, and another pupil, Volcher Coiter, who taught anatomy at Bologna before becoming surgeon to the German army in Nuremberg.

3 *Anat. Lect.*, pp. 255, 259, 263, 267.

4 *Ibid.*, pp. 275–99.

5 To this point, this passage is a direct quotation form Bauhin's *Theatrum anatomicum*, II, 24. Bauhin continues 'and is carried to the lungs', an anatomical error. The foramen ovale is nowadays described as situated between the right and left atria.

6 Bauhin's phrase. In the *Anat. Lect.*, p. 263, Harvey says that it is a valve. For his observations there on the foetal heart, see pp. 127, 251, 263.

7 The 1628 text is corrupt in this passage. It reads:

> *Haec, inquam, membrana sic constituta est, ut dum laxe in se concidit, facile ad pulmones et cor via resupinetur, et sanguini a cava affluenti cedat quidem; at, ne rursus in cavam refluat, impediat. . . .*

It is clear from what Harvey has just said that he knows that the blood from the vena cava goes through the foramen ovale directly into the left atrium. The statement therefore, *facile ad pulmones et cor via resupinetur* contains an anatomical error. Furthermore, *via resupinetur*, literally 'a way is turned on its back', does not make sense.

The 1648 text, aware of the difficulty, has tried to emend the reading but has only made matters worse:

> *Haec . . . ut, dum laxe in se concidit, faciat ad pulmones et cor viam, resupinetur*, etc.

In 1653, this was translated:

> This membrane . . . is so contrived . . . that hanging loosely with its own weight, it makes way into the lungs and heart, and is turned up, giving passage to the blood which flows from the vena cava. . . .

A footnote has been added to the Latin text and to the translation saying that this membrane is the septum. But this it is clearly not.

Willis, using the 1628 text, has disregarded the anatomical error and mistranslated to make sense:

> This membrane . . . is so contrived . . . that falling loosely upon itself it permits a ready access to the lungs and heart, yielding a passage to the blood . . . (p. 37).

K. J. Franklin has done no better: 'falling back loosely on itself, it moves easily in the direction of the lungs and heart . . . (p. 46).

As I pointed out in note 5 above, Harvey has taken his description of the foramen ovale almost verbatim from Bauhin's *Theatrum anatomicum*. When he comes to this membrane, Bauhin writes:

> *At ne sanguis in Cavam refluat, hujus foraminis regioni, quae Arteriae venosae cavitatem respicit, praefecta est operculi instar membrana . . . quae ad solam radicem firmatur, reliquo toto corpore in vasis cavitate pendula; quo dum laxe in se concidit, facile ad pulmonis vas resupinetur, et sanguini a Cava impetu affluenti, cedat quidem; at ne rursus in Cavam refluat, impediat.*

There would seem to be little doubt that here we have the source of Harvey's sentence, and that the text which the printer corrupted originally read:

> *Haec, inquam, membrana sic constituta est ut, dum laxe in se concidit, facile ad pulmonis vas et cor resupinetur, et sanguini. . . .*

The phrase *pulmonis vas*, 'the pulmonary vessel', refers to the pulmonary vein, that is the left atrium, and *cor* to the left ventricle of the heart. The printer misread Harvey's abbreviation of *pulmonis* as *pulmones*, misread or misunderstood *vas*, and supplied *via* in its place and probably altered the order of the words. My translation here follows my suggested emendation.

8 As the foetal heart was not supposed to beat, it was thought that both the venous and arterial blood derived directly from the mother through the umbilical vessels and ebbed and flowed to and from the maternal heart. The coronary vein and artery provided the heart with the means to exist. Accordingly, no blood was believed to be present in the right ventricle of the foetal heart and no spirit in the left. See Fabricius, *De formato foetu*, II, 2 and 8.

9 *De spiritu*, cap. 4, 483a 14.

10 Cf. *Anat. Lect.*, p. 263: 'Both these passages remain open after weaning and in some animals which are not perfect they are even always so, as dove, goose.'

11 Botallus described this 'vein' connecting the right and left ventricles in a small tract published from Paris in 1565, *Vena arteriarum nutrix a nullo antea notata.* His contemporaries quickly pointed out that he was describing the abnormal persistence of the ductus arteriosus and this was known to Galen.

12 Harvey is not known to have written this treatise and no manuscript notes for it are known to exist. A considerable number of observations on the lungs and on respiration will be found in the *Anatomical Lectures, q.v.* Introduction, pp. liii–lvi.

CHAPTER 7

That the blood does pass from the right ventricle of the heart through the parenchyma of the lungs into the pulmonary vein and left ventricle of the heart

It is well enough known that this can be done and that there is nothing to prevent it, when we consider how water passing through the substance of the earth brings forth rivers and springs, or observe how sweat passes through the skin, or how urine flows through the parenchyma of the kidneys. It is to be noticed that those who drink the waters of Spa, or those of Our Lady (*della Madonna*, as they call them) in the Paduan countryside,[1] or other mineral or sulphurous waters, or those who in carousing swill themselves with drink, in an hour or two piss out the whole of it through their bladder. This great quantity must stay for a while to be concocted, must flow through the liver (as everyone agrees the juice of the food we eat does twice every day), must flow through the veins, through the parenchyma of the kidneys and through the ureters into the bladder.

What men are these then, whom I hear denying that the blood, indeed the whole mass of the blood, can pass through the substance of the lungs quite as well as the nutritive juice through the liver, deeming it a thing impossible and in no wise to be believed? They are the kind of men who, in the Poet's words,[2] agree that a thing may assuredly be so when they wish it so, but when they do not wish it, then by no means can it be.

Here where there is need, they are afraid to assert it, but where there is no need, they have no such fear.

The parenchyma of the liver and that of the kidneys likewise is denser by far than that of the lungs, which is of a much finer texture and spongy in comparison with that of the kidneys and liver.

In the liver there is nothing that impels, no driving force; in the lungs, the blood is kicked on by the pulse of the right ventricle of the heart and by this impulsion of the blood the vessels and porosities of the lungs needs must be distended. Besides, as Galen says in *De usu partium*, in respiration the lungs rise and fall, by which motion it follows of necessity that their porosities and their vessels are opened and shut, just as it happens in sponges and in all other things having the constitution of a sponge, when they are squeezed and then expand again. The liver, on the contrary, lies still and has never been seen to be expanded or constricted in this way.

Lastly, since there is no one but affirms that the juice of all the food that is eaten can pass through the liver into the vena cava, as well in man as in an ox or in the largest animals, and they are all forced to affirm it because, for there to be any nutrition, the food must pass some way and go into the veins and there is no other way but this, why then should they not likewise believe this of the passage of the blood through the lungs in men come to age, upon the same arguments? And with Columbus, a most skilful and learned anatomist, believe and assert the same from the largeness and structure of the vessels of the lungs, and because the pulmonary vein and likewise the ventricle are always full of blood which must needs have come hither out of the veins and by no other path but through the lungs, as both he and I from the things which we have already said, from our own eyesight and from other arguments, believe to be most plain.[3]

But seeing that there are some persons who admit of nothing unless there be an authority alleged for it, let them know that this truth can be confirmed even from Galen's own words, that is to say, not only that the blood can be transmitted from the pulmonary artery into the pulmonary vein and then into the left ventricle of the heart and afterwards into the arteries, but also that this is done by the incessant beating of the heart and the movement of the lungs whilst we breathe.

There are in the orifice of the pulmonary artery, three doors, made like a sigma or half-moon, which altogether prevent the blood sent into the pulmonary artery from returning again to the heart. All men know of the need for these valves and of their use. Galen explains it in *De usu partium*, Bk VI, cap. 10, in these words and says:[4]

> There is in general, a mutual anastomosis and opening of tiny mouths between the arteries and the veins, and they borrow blood and spirits alike from one another by invisible and very narrow passages. But if the mouth of the arterial vein always stood open and Nature had found no device to shut it when requisite and to open it again, it could never have come to pass that, the thorax being contracted, the blood should be carried over into the arteries through those invisible and very tiny openings. For everything is not attracted out of something else, nor sent out from it, in the same way. For just as a light thing is more easily attracted than something which is heavier when the instruments for so doing are enlarged, and expelled when they are contracted, so anything is more quickly attracted in and sent forth again through a wide passage than through a narrow one. When the thorax is contracted, the venous arteries which are in the lungs, being pushed and compressed together strongly on every side, squeeze out as quickly as possible the spirit that is in them; but they take

> in through those fine openings a part of the blood, and this would indeed never happen if the blood could flow backwards into the heart through that very large opening (such as is that of the arterial vein) into the heart. Now, therefore, its return through that great opening being stopped, when the blood is compressed on all sides, a little of it distills into the arteries through those tiny orifices.

A little later in the following chapter, Galen says:

> The more the thorax strives vehemently in squeezing out the blood, the more precisely do those membranes (that is to say the sigmoid valves) close its mouth and prevent anything from flowing back.

And again, in the same Chapter 10, a little before, he says:

> Unless there were valves, a threefold inconvenience would follow, namely, that the blood itself would continually travel over this long course to no purpose, flowing forwards in the diastole of the lungs and filling all the veins in them, and in systole, like a tide of the sea, flowing back again, hither and thither with a movement like that of Euripus, and this would by no means be convenient for the blood. However, this may seem no great matter. But that in the meantime it should weaken the benefit of respiration, this would no longer be considered a mere trifle . . .

and so forth. A little further on:

> And a third inconvenience would also follow and no slight one, seeing that the blood would flow backwards at every expiration, unless our Maker had fashioned that outgrowth of membranes.

From this he concludes in chapter 11:

> The common use of all these (that is of the valves), is to prevent the backward return of any matter and each of them has a proper use, some to draw materials from the

> heart and prevent them from returning to it again, the others to draw materials into the heart and prevent them from flowing out again. For Nature would not have the heart to be wearied with needless labour, nor to send forth anything at any time to any part from whence it had been better to extract it, nor extract again from that part to which there was need to send. Therefore, since there are altogether four orifices, two in each ventricle, one of these leads in and the other leads out.

And a little after:

> Furthermore, seeing that there is one vessel consisting of a single coat implanted in the heart, and the other consisting of a double coat coming forth from it, a place common to both, that is to say the right ventricle (thus Galen, but for the same reason likewise I understand the left ventricle of the heart), had of necessity to be prepared to be a kind of pool to which both belonged that the blood might be drawn forth from it by the one and sent into it by the other.

This argument which Galen puts forward in support of the passage of the blood from the vena cava through the right ventricle into the lungs, we may more rightly use for the passage of the blood from the veins through the heart into the arteries, changing only the names. It clearly appears, therefore, from the words and arguments of Galen, that great man and the father of all physicians, that the blood passes from the pulmonary artery through the lungs and into the branches of the pulmonary vein both by reason of the pulsation of the heart and because of the movement of the lungs and thorax. (See the Commentary of the learned Hofmann on Galen's *De usu partium*, Bk VI. I saw this book only after I had written these things.[5])

From this it appears also that the heart receives the blood

continually into the ventricles as into a pool and sends it forth again, and for this reason it was necessary that it should be provided with four sets of valves, two of which serve for the entering in of the blood and two for its expulsion, lest the blood should either be inconveniently driven hither and thither, like the sea in the channel of Euripus, or flow back to the place from whence it were fitter it should be drawn, or away from that place to which it was needful that it should be sent, and so the heart be wearied with vain toil and the respiration of the lungs be hampered. And lastly it appears that my assertion is plainly true, that the blood flows continuously and incessantly through the porosities of the lungs out of the right ventricle into the left, out of the vena cava into the great artery; for seeing that the blood is continuously sent from the right ventricle into the lungs through the pulmonary artery and likewise is continuously drawn from the lungs into the left ventricle, as is clear from what has been said and from the siting of the valves, it is not possible for it to do otherwise than to pass through continuously.

And likewise, seeing that always and without intermission the blood enters into the right ventricle of the heart and continuously goes forth from the left (which is clear alike from the argument and from observation), it is impossible for the blood to do otherwise than pass across incessantly from the vena cava into the aorta.

Therefore, I maintain that that which is done through most open ways in the greater number of animals and in all while they are still growing, as is plainly to be seen from dissection, is accomplished in these animals when they are grown through hidden porosities in the lungs and the tiny holes in their vessels, and that this is equally clear both from Galen's words and from what I have already said. From this it follows that although one ventricle of the heart, namely the left, were sufficient for the distribution of blood throughout the whole body and for its drawing forth from the vena cava, just as indeed it does happen in all those creatures that have no lungs, yet Nature, desiring the blood itself to be percolated through the lungs, was forced to add the right ventricle by whose pulsation the blood should be driven through the very lungs out of the vena cava into the receptacle of the left ventricle. And in this sense only it is to be said that the right ventricle was made for the sake of the lungs and on account of the transference of the blood and not for the sake of nutrition. For it is altogether inconsistent with reason to suppose that the lungs should have a greater need of such an abundance of food and that supplied by force and so much more pure and full of spirit, as being immediately conveyed from the ventricles of the heart, than either the most pure substance of the brain or the most resplendent and divine constitution of the eyes or the flesh of the heart itself, which is more directly nourished through the coronary artery.

NOTES

1 Medical students at Padua were generally taken during the months of the summer vacation to Abano and other places in the Euganean hills to study the effects of drinking the mineral waters from the various springs and to learn their use in the practice of medicine. In the *Anatomical Lectures*, Harvey discusses the formation of urine and its passage from the kidneys to the bladder.

2 Unidentified.

3 *De re anatomica*, Bk VI; Bk XI, cap. 2. Harvey discussed Columbus's findings at length in his *Anatomical Lectures*.

4 When Galen upset the neat system which maintained that air, or spirit, was contained in the arteries and blood in the veins by proving that the arteries contained blood, it was found necessary in order to explain how it

got there to invent the existence of anastomoses between veins and arteries so that the blood could pass into the arteries and the spirit enter the veins. So Harvey here begins with the Galenic statement concerning these anastomoses in the lungs. His understanding of them, however, was not in accordance with Galenic theory. His acceptance of the necessity for the pulmonary transit of the blood compelled him to postulate the existence of a communication between arteries and veins but he did not, of course, believe that the blood could pass indifferently from arteries to veins and veins to arteries. He returns to the question of anastomoses again in Chapter 9, p. 83. Here it will be noticed that in his summary of the passages quoted from Galen, Harvey omits all reference to anastomoses, for his intention is merely to stress Galen's understanding of the competence of the valves of the heart. Scarcely anyone before Harvey, except Realdus Columbus, understood the significance of the competence of these valves. Harvey's own knowledge of their action is revealed in Chapter 17 and it is most probable that it was only because he was already convinced by his own observations of the true manner of their action and of their competence that he could turn back to these passages from Galen and point out that here, in fact, Galen had described their action correctly.

As the terms pulmonary vein and pulmonary artery are not easily applicable in the Galenic context, I have here adopted the old but confusing terminology of venous artery and arterial vein respectively.

5 Caspar Hofmann, *Commentarii in Galeni de usu partium corporis humani*, Frankfurt 1625, pp. 107, 120–1.

CHAPTER 8

Of the abundance of blood passing through the heart out of the veins into the arteries, and of the circular motion of the blood

Thus far then there are, perhaps some who will say that they agree with me about the passing of the blood out of the veins into the arteries, and about the ways through which it goes, and how it is sent on and distributed by the beating of the heart, having been persuaded of it already either by the authority of Galen, or by the arguments advanced by Columbus and by others. But now what remains to be said about the abundance of this blood which passes through, and where it comes from, though they be things most worthy of consideration, when I shall say them, so new and unheard of are they, that not only do I fear some mischief to myself from the malice of certain persons,[1] but I am likewise afraid lest I have all men to my enemies, so much does custom and doctrine once absorbed and deeply rooted, like second nature, weigh with all men and the revered appearance of antiquity constrain them. However, the die is now cast and I put my trust in the love of truth and in the integrity of learned minds.

Now truly,[2] when I had many times and seriously considered with myself the varied means of searching, and how many there were! both from the dissection of living creatures for experiment's sake, and from the opening of arteries, as well as from the symmetry and great size of the ventricles of the heart and of the vessels which go into it and go out from it (for Nature who makes nothing in vain would not have allotted to those vessels so comparatively large a size to no purpose), and from the carefully balanced and exquisite contrivance of the valves and fibres and from the rest of the fabric of the heart, and from many other things, and when I had for a great while turned over in my mind these questions, namely, how great was the abundance of the blood that was passed through and in how short a time that transmission was done, and when I had perceived that the juice of the food that had been eaten could not suffice to supply the amount of the blood—nay, more, we would have veins empty and altogether drained dry and arteries, on the other hand, burst open with the too great inthrusting of blood, unless this blood should somehow flow back out of the arteries once more into the veins and return to the right ventricle of the heart—I began to bethink myself whether it might not have a kind of movement as it were in a circle. And this I afterwards found to be true, and that the blood is thrust forth and driven out of the heart through the arteries into the whole structure of the body and into all its parts by the pulsation of the left ventricle of the heart, just as it is driven into the lungs through the pulmonary artery by the beating of the right ventricle, and that it flows back again through the veins into the vena cava and into the right auricle, just as it returns from the lungs through the pulmonary vein into the left ventricle of the heart as I have already explained.

We may call this motion circular in the same way in which Aristotle says that the air and the rain imitate the circular motion of the heavens. For the earth being wet evaporates by the heat of the sun; the vapours being drawn upwards condense and being condensed descend again in raindrops and wet the earth. In this way here on earth all generation is accomplished, and so too from the circular movement of the sun, its approach and recession, proceed the beginnings of storms and meteors.[3]

So in all likelihood it happens in the body that by the movement of the blood all the parts are fed and nourished, fostered and quickened, by blood that is warmer, perfected, vaporous, full of spirit and, so to speak, alimentative; that on the other hand in the parts the blood becomes cold and coagulated and as it were enfeebled, whence it returns to its beginning, the heart, as to the fountain or inmost shrine of the body to recover its perfection.[4] There again by natural heat, powerful and vehement, as it were the treasure of life, it is again made liquid and is dispensed again from thence through the body, fraught with spirits, as with balm,[5] and these things depend upon the motion and pulsation of the heart.

So is the heart the first principle of life and the sun of the microcosm, just as the sun deserves equally to be called the heart of the world,[6] by whose virtue and pulsation the blood is moved, made perfect, quickened and preserved from corruption and lumpiness, and this familiar household god performs his office for the whole body by nourishing, cherishing and quickening, being the foundation of life and the author of all things. But of these things it will be more appropriate to speak when we come to enquire of the final cause of this kind of movement.

Hence it is that since the veins are certain ways and vessels carrying blood, there are two kinds of blood vessels, the vena cava and the aorta, not by reason of their position on this side and on that of the heart, as Aristotle thought,[7] but by reason of their office, and not as is commonly supposed by reason of their composition, seeing that in many creatures, as I have said, a vein does not differ from an artery by the thickness of its coat, but they are differenced by their office and by their use. A vein and an artery, as Galen observed, were both called veins by the Ancients and not unjustly, because the latter, the artery, is a vessel carrying blood away from the heart into the whole of the body, while the former, the vein, is a vessel bringing the blood back again from the body into the heart; the artery is the way from the heart, the vein the way to it. The vein contains blood that is more crude, spent and now become unfit for nutrition; the artery, blood that is concocted, perfected and alimentative.

NOTES

1 This strange remark is reminiscent of what Harvey had already said at the end of Chapter 1, p. 30. We do not know whom he had in mind.

2 A discussion of the text and translation of this passage will be found above, Introduction, Section 3.

3 This paragraph is a composite affair made up of many statements derived from different passages in Aristotle's writings, see *De mundo*, chapter 6, 399a 20–35; *De anima*, II, 4, 415b 3–8; and more generally, *De generatione et corruptione*, II, 10. See also Harvey's *De generatione*, Ex. XXVIII.

4 All the ideas expressed here on the heart and blood derive from Aristotle and will mostly be found easily in *De partibus animalium*. On heat and cold as the sources of movement see his *De motu animalium*; cf. also Harvey's *De motu locali animalium*, p. 25.

5 'Balm' or 'balsam', meaning an unguent or a healing or soothing agent, was frequently used in the seventeenth century with the extended connotation of a 'preservative'. There is no reason to suppose that it is here used by Harvey in any other sense.

6 These are all commonplace notions to seventeenth-century writers and derive from Aristotle's views on the supremacy of the heart.

7 *P.A.* 666b 30; 667b 32–668a 4.

CHAPTER 9

From the proof of the first proposition it is shown that there is a circulation of the blood

But lest anyone should think that I am speaking empty words and making fair assertions only without foundation, and introducing a new thing without just cause, there are three propositions which come now to be proved and when they have been established I think that this truth must needs follow and the matter be apparent to all men.

First, that by the pulsation of the heart the blood is continuously and incessantly transferred out of the vena cava into the arteries in so great a quantity that it cannot be provided by the food we eat, and in such a way that the whole mass of the blood passes across from thence in a short time.

Second, that by the pulse of the arteries the blood is continuously, consistently and incessantly driven and enters into every member and part of the body in far greater quantity than is necessary for nutrition or can be supplied by the total amount of the blood.

Third, that in the same way the veins themselves bring back this blood uninterruptedly from each member into the place of the heart.

These propositions being proved, I think it will be abundantly clear that the blood goes around and is returned, is driven onwards and flows back, out of the heart into the extremities and from the extremities back into the heart again, and thus completes a movement as it were in a circle.

Let us suppose, either theoretically or experimentally, how much blood the left ventricle may contain in its dilatation,[1] that is when it is full, say either two ounces, or three ounces or one and a half. I have found in a dead man over two ounces.

Let us suppose likewise how much less the heart may contain in its contraction, or how much the heart contracts itself, and how much smaller a capacity the ventricle has in its contraction or contractions, and how much blood is thrust out into the great artery, for in the systole there is always some blood thrust out as I proved before in Chapter 3, and all men acknowledge that it is so, convinced by the structure of the valves. Let us put up as a reasonable guess that a fourth, fifth or sixth part, and at least an eighth, is sent into the artery.

Therefore, let us suppose that in a man there is sent forth at every beat of the heart half an ounce or three drams or one dram which cannot possibly return to the heart by reason of the hindrance of the valves.

The heart in one half hour makes above a thousand pulses; in some men, indeed, and at some times, two, three or four thousand. Now multiply the drams and you will see that in one half hour either one thousand times three drams or two drams, of five times one-hundred ounces, or some such proportionate quantity of blood, is passed through the heart into the arteries, that is, always in a greater quantity than can be found in the whole of the body. Similarly, in a sheep or a dog, if it be but one scruple that passes through in one contraction of the heart, then in one half hour a thousand scruples, or about three and a half pounds of blood will have passed through, but in their bodies there is generally not contained more than four pounds of blood, for I have tried it in a sheep.

So, having calculated according to the least amount of blood that we may suppose to be transmitted and multiplied it by the number of pulsations, it would seem that the whole quantity of the mass of blood passes through the heart from the veins into the arteries, and likewise through the lungs.

But, grant that it be not done in one half hour, but in one hour or in one day, whatever you choose, it is clearly evident that more blood is incessantly driven through the heart by its pulsation than can possibly be either supplied by the food we eat or contained at one time in the veins.

And it must not be said that the heart in its contraction thrusts out blood sometimes and sometimes does not, or thrusts out practically nothing or some imagined thing, for this has already been refuted, and besides, it is contrary to observation and reason. For if in the dilatation of the heart it needs must be that the ventricles are filled with blood, it is likewise necessary that in its contraction it must always thrust forth blood and that not a little, for the vessels are not small nor is the heart's contraction slight. In whatsoever proportion, be it a third, a sixth or an eighth, that the blood is thrust out, that blood had to be previously contained in the heart and filling it in its dilatation, so that the capacity of the contracted ventricle is in the same proportion to its capacity when dilated. And because in its dilatation it does not happen that it is filled with nothing or with some imaginary thing, so in its contraction it never expels nothing or that which is imaginary, but always an amount proportionate to the contraction. Therefore it is to be concluded that if in a man, a sheep or an ox, the heart in one pulsation sends out one dram of blood, and if in one half hour there are one thousand pulses, it happens that in the same time ten pounds and five ounces of blood are transmitted. If at one pulse it sends out two drams, in half an hour it would send out twenty pounds and ten ounces; if half an ounce, then forty-one pounds, eight ounces. If one ounce, then eighty-three pounds four ounces will have been transferred from the veins into the arteries, I repeat, in one half hour.

But how much is actually thrust out at each pulsation, and when more and when less and for what reason, I will perhaps explain more accurately later from the many observations which I have made.[2]

Meantime of this I am certain and would wish all men to bear it in mind, that sometimes the blood flows across in a more abundant quantity, sometimes in a less, and that sometimes the circuit of the blood is accomplished more quickly and sometimes more slowly, in accordance with temperament, age, external and internal causes, natural or non-natural occurrences, sleep, rest, food, exercise, passions of the mind and so forth.

But indeed, even if the blood passes through the lungs and heart in the smallest quantity that may be, a far greater abundance of it is conveyed into the arteries and the whole body than can possibly be supplied by the food we eat or in any other way whatsoever, unless it return again round a circuitous course.

This also is clearly to be seen by any who watch the dissection of living creatures, for not only if the great artery be cut, but, as Galen proves, even in man himself, if any artery even the smallest be cut, in the space of about half an hour the whole mass of the blood will be drained out of the whole body, as well out of the veins as out of the arteries.

Butchers also can testify to all these things well enough, when in slaughtering an ox they cut the carotid arteries and in less than a quarter of an hour drain out the whole mass of blood and empty all the vessels. We find the same thing happening again, sometimes in a very short time, in the amputation of limbs or the excision of tumours, as a result of the enormous profusion of blood.

Nor does any man blunt the force of this argument by saying that in the slaughtering of animals and amputation of limbs, the blood flows out as much through the veins as through the arteries if not more, seeing that indeed the matter is quite the reverse. The veins shed but very little blood because they flap down, because in them is no force driving the blood out, and because the position of the valves in them (as shall hereafter appear) obstructs the blood, whereas the arteries pour out the blood driven by a propelling force in very great quantity and in spurts, as if it were ejected out of a syringe.[3] But the matter can easily be tried. Leave the vein untouched and cut the carotid artery in a sheep or a dog and you will see with amazement with what great force, in how great profusion and how quickly it happens that all the blood empties out from the whole body, as well from the veins as from the arteries.

It is quite clear from what has been said already that the arteries receive blood from the veins nowhere except by means of the transference made through the heart. If you then ligate the aorta at the base of the heart and open the carotid or any other artery, and if you then see that the arteries alone are empty and the veins full, you will no longer have any cause to doubt.

From this you will clearly see the reason why in an Antomy so much blood is found in the veins and so little in the arteries, why there is much in the right ventricle and little in the left, a fact which perhaps gave the Ancients occasion to doubt and to think that spirits alone were contained in the hollow channels of the arteries while the animal was alive. But the real reason is perhaps because there is nowhere any passage afforded to the blood from the veins into the arteries except through the heart itself and through the lungs. When the lungs have died and cease to move, the blood is prevented from passing from the little branches of the pulmonary artery into the pulmonary vein and from thence into the left ventricle of the heart, just as it was observed before that in an embryo it was prevented from so doing by reason of the lack of movement of the lungs to open and shut the tiny holes and hidden and invisible porosities. But since the heart does not cease to move at the same time as the lungs, but continues to beat afterwards and to outlive them, it follows that the left ventricle and the arteries send forth blood into the veins and into every part of the body, but receive none through the lungs and therefore they are as it were empty.[4] And this also brings no little credibility to my thesis, for no other reason can be adduced to support it except that which I have maintained in my hypothesis.

Furthermore it is clear from this that the more strongly and the more vehemently the arteries beat, the more quickly will the body be emptied of all the blood in any case of haemorrhage. Hence also it follows that in any fainting fit, moment of fear or the like, when the heart beats more languidly and weakly, without any force, every haemorrhage is quietened and arrested.

Again, it also follows from this that in a dead body when the heart has ceased to beat, you cannot, try as you will, extract more than half of the whole mass of the blood from the jugular or crural veins or arteries when they are opened. Nor can a butcher after he has knocked the ox on the head and stunned him, draw out all the blood from his body unless he has first cut his throat before the heart has ceased to beat.

Finally, from this we may have some opinion concerning the anastomosis of veins and arteries, where it occurs and how it is made and why it is that no one till now has said anything about it that is correct. That enquiry I am now pursuing.[5]

NOTES

1 The first person to attempt to apply mathematical calculations to biological problems seems to have been Sanctorius Sanctorius (1561–1636). He conceived the idea of living in a specially built balance in order to study his

variations in weight and in particular to find out the amount of 'invisible transpiration' through the skin. He pursued this investigation over many years and published its results in the form of a series of aphorisms called *De statica medica*, printed in Venice in 1614. I cannot find the slightest resemblance between this work and Harvey's calculations.

Harvey's argument from quantity is the first of its kind in the history of physiology, and though the estimates are very low, the conclusions are irrefutable. It is now known that the heart expels about 70 cm^3 per beat in resting conditions and that is the equivalent of $2\frac{1}{2}$ oz in Harvey's measure. He uses apothecaries or troy weight in which 3 scruples = 1 dram; 8 drams = 1 oz; 12 oz = 1 lb. In resting conditions, the heart beats 70 times per minute so that the total output of the heart is about 5 l per minute.

2 Harvey did not publish these observations and they are not now known to exist.

3 The Latin is *sypho*. Cf. the Second Letter to Riolan: '. . . just as you might feel in the palm of your hand water squirted through a syringe (*sipho*) at diverse and several shootings' (1653 text, p. 149). It is only later in the same Letter (pp. 183–4) that he compares the jetting of arterial blood with water driven through pipes by an engine and imitating the strokes of the pump. In this case *sipho* is used with the meaning of fire-engine. But in *De motu cordis*, this analogy of the heart's action with a pump had not yet occurred to Harvey. See *WH and the Circulation*, pp. 169–72.

4 Harvey is describing the appearance of the heart after death from asphyxia. Cf. his remarks in the Second Letter to Riolan on the enormously distended right auricle in a body he dissected two hours after death by hanging. Bodies dissected at formal Anatomies were those of hanged criminals. As the p.m. appearance of the heart varies with the cause of death, this description does not contradict his remarks on the presence of blood in both ventricles in chs 5 & 17, where the cause of death is unknown.

5 Presumably a reference to Harvey's attempt to find the actual connections between veins and arteries. For his later discussions of this problem see his Letter to Hofmann 1636, his Second Letter to Riolan, 1649 and his Letter to Slegel 1651. As is well known the actual capillary connections were first seen by Marcello Malpighi and their description published by him in 1661.

CHAPTER 10

The first proposition concerning the quantity of blood passing from the veins into the arteries and concerning the existence of the circulation of the blood is defended from objections and proved by further experiments

Thus far then the first proposition has been proved both by an appeal to calculation and by reference to experimental evidence and actual observation, that is to say, the proposition that the blood passes into the arteries continuously and in greater abundance than can be furnished by the food eaten and in such a way, seeing that the whole mass of the blood passes through in so short a time, that it needs must be that the blood makes a circuit and returns.

But suppose someone should say that the blood can indeed pass through in great abundance, but that it is not necessary for it to make a circuit, seeing that it may happen that it is replenished from the food that is eaten, and that the provision of milk in the paps affords an example of this; for a cow in one day gives three, four or seven or more gallons of milk, and a woman when suckling one child or two produces every day two or three pints, and this is clearly replaced by the food that is eaten. To this it must be answered that the heart is known to send out as much or more in one hour or two according to the calculation that has been made.

But suppose that he is not yet convinced and should insist further, saying that although it may happen that the blood is poured forth violently when an artery is cut and a way opened that is contrary to Nature, yet it does not so happen in an intact body and when no exit is provided and the arteries are full or in their natural condition that so great an abundance passes through in so short a time that it needs must be that a return is made by the blood. To this it must be answered that it is evident from the aforesaid reckoning and from the argument derived from it, that as much more as the heart contains when it is filled in its dilatation than it contains in its contraction, so much is sent out, for the most part, at each pulsation, and therefore it passes through in such great abundance in the intact body, constituted according to Nature.

But in snakes and in some fish, if you ligate the veins a little below the heart, you will see the space between the ligature and the heart very quickly emptied, so that you must needs affirm the return of the blood, unless you would deny your own eyesight. This same fact will also be clearly obvious later in the proof of my second proposition.

I will conclude by proving all these things from one example to which everyone can give credit on the testimony of his own eyes. If he will cut up a live snake, he will see the heart beat calmly and distinctly for more than a whole hour and contract itself in length like a worm in its constriction (for the snake's heart is elongated), and thrust itself out again. And he will see that in its systole it is paler in colour and darker in diastole. And he will see almost all the other things by which I have said that this truth would be manifestly proved, for here in the snake all these things take place far more slowly and distinctly.

Now this one experiment in particular can be tried and it is more illuminating than the noonday sun. The vena cava enters the lower part of the heart; the artery comes out at the upper part. Now, taking hold of the vena cava with a pair of forceps, or with your finger and thumb, and the course of the blood being stopped a little way below the heart, you will see almost immediately after the heartbeat that the space between your fingers and the heart is emptied, the blood having been drained out by the pulse of the heart; and at the same time the heart will be of a much lighter colour even in its dilatation and smaller too for want of blood, and at last it will beat more faintly so that finally it seems like to die. But immediately, as soon as you let go the vein, both its wonted colour and size return to the heart. Afterwards, if you relinquish the vein and ligate or compress the artery in the same way at a little distance from the heart, you will on the contrary see the parts where they are compressed swell vehemently, and that the heart is swelled beyond measure and acquires a purple colour to the point of being blueish-black, and that it is at last so oppressed with blood that you would think it would be suffocated. But, letting go the tie, you will see that the heart returns again to its normal constitution in colour, size and beat.

So then here are two kinds of death,[1] extinction by reason of deficiency and suffocation by too great quantity, and here you may have the example of both before your eyes and confirm the truth which has been spoken by your own observation of the heart.

NOTES

1 Cf. Aristotle, *De iuventute et senectute*, caps. 4 and 5. While heat remains in the heart, life remains in the body. Death is the destruction of this heat and can be brought about either by exhaustion or extinction. Extinction results from lack of supplies; exhaustion from the excessive accumulation of heat from lack of refrigeration when respiration ceases.

CHAPTER 11

The second proposition is proved

That the second proposition that I have to prove may appear more clearly to those who are considering the matter, some experiments are to be taken note of which show plainly that the blood enters every member of the body through the arteries and returns through the veins, and that the arteries are the vessels which carry the blood away from the heart and that the veins are the vessels and ways by which the blood returns to the heart itself; and that in the members and in the extremities the blood passes from the arteries into the veins either directly by means of an anastomosis or indirectly through the porosities of the flesh, or by both these means, just as it has already been shown that it passes from the veins into the arteries in the heart and in the lungs.[1] From these things it will be plainly evident that the blood is moved in a circuitous course from thence hither and from hence thither, that is to say, from the centre into the extremities and from the extremities back again to the centre.

Furthermore, if afterwards the reckoning again be made as before, what was then said about the abundance of the blood will be evident, namely that it cannot be supplied by the food eaten, nor is so great an abundance required of necessity for nutrition.

At the same time some matters which concern ligatures will also become clear, as to why they cause an afflux of blood which is not to be explained by heat or pain or by the attractive force of a vacuum or by any other cause previously known, and also as to the convenience and usefulness which ligatures afford in the practice of medicine, and how they arrest and provoke a haemorrhage, and why they produce gangrene and necrosis of the members and thus are of use in the gelding of some animals and in the excision of fleshy tumours and warts. For indeed, because no one has rightly understood the causes and logical basis for these ligatures, it has come about that nearly all recommend and advise their use in the treatment of diseases according to the opinion of the Ancients, but there are few indeed who by their proper application afford any relief to those whom they treat.

Ligatures are either tight or middling tight. By a tight ligature I mean one in which the limb is so tightly bound all round by a narrow band or a noose that you cannot see the arteries beat anywhere beyond the ligature; such a one as we use in the amputation of limbs when we foresee an efflux of blood, or again in the gelding of animals and the excision of tumours, for by it the accession of nutriment and heat is entirely interrupted and we see the testicles and great sarcoses wither and die away and afterwards drop off.[2]

By a middling-tight ligature I mean one in which the limb is compressed all round but without causing pain, and such that it allows the arteries to beat a little beyond it; such a one as is used in the drawing and letting of blood, for though the ligature be made above the elbow, yet you can feel the arteries beat a little in the wrist if the ligature be properly made for phlebotomy.

Now let there be an experiment made on a man's arm either by employing a narrow band such as they use in blood-letting or by compressing the arm rather firmly with the hand; and this is more conveniently done in a lean man and one whose veins are rather large and when the extremities are warm, the body having been heated by exercise, and a greater quantity of blood is in the extremities and the pulsation more violent, for then everything will appear more clearly.

If you then make a tight ligature, drawing it as hard as it can be endured, you may observe first that beyond the ligature, that is towards the hand, the artery does not beat either in the wrist or anywhere else. Next, that the artery above the ligature immediately swells higher in its diastole and beats more and higher and more vehemently, and beside the ligature itself, swells with a kind of tide as if the blood did strive to burst open the barrier to its flow and unlock the passage which is stopped, and that the artery appears to be fuller there than normal. Lastly, you will see that the hand will retain its usual colour and condition and only in the course of time begin to grow somewhat cold, and that nothing whatsoever is drawn into it.

After this ligature has continued a while, let it be suddenly loosened into a ligature of middling tightness, such as I said they use in the letting of blood. It is to be observed that the whole hand is immediately imbued with colour and distended, and that its veins become swollen and lumpy, and in the space of ten or twelve pulses of the artery you will see that the hand is exceedingly full with the much blood that is driven and forced into it and that a great quantity of blood is quickly drawn by that middling-tight ligature without either pain or heat or the flight from a vacuum or any other cause mentioned above.

If at the very moment of the slackening of the ligature, any one will carefully put his finger on the pulsating artery beside the ligature, he will feel the blood as it were slipping underneath his finger.

Moreover, he on whose arm the experiment is made, from the moment of the loosening of the ligature from tight to middling tight, will plainly feel the heat and blood entering the arm below it in time with the pulse, as if some obstacle had been removed, and he will have the sensation of something passing along the course of the arteries and suddenly dilating them and distributing itself here and there through his hand, and immediately his hand will become hot and distended.

Just as in a tight ligature the arteries above it are distended and pulsate and those below do not, so on the contrary in a middling-tight ligature the veins below it swell and become resistant but not so those above, and the arteries below it become smaller. Indeed, if you were to compress those swollen veins, unless you did it very strongly, you would scarcely see either the blood being diffused or the veins distended above the ligature.

So from these things it is easy for any man that will observe carefully to know that the blood enters by the arteries, for when they are tightly ligated nothing is drawn by this ligature, the hand keeps its colour, nothing flows into it and no distension happens. But when the arteries are a little freed, as in a middling-tight ligature, it is obvious that the blood by force and impulsion is abundantly thrust in, and the hand becomes swollen. That is to say that where the arteries pulsate as they do in the hand in the case of a middling-tight ligature, the blood flows forward; where they do not pulsate, as in the case of a tight ligature, the blood does not flow at all, except above the ligature. Meantime, when the veins are compressed, nothing can flow through them. The clear indication of this is that below the ligature the veins become much more swollen than above it and than they are wont to be when the ligature is removed, and that when they are compressed they convey nothing to their upper parts. And hence it is clearly manifest that the ligature impedes the return of the blood through the veins into the superior parts and makes those beneath the ligature to remain swollen.

But for good reason the arteries, notwithstanding the middling-tight ligature, thrust out the blood by the strength and impulsion of the heart from the inner parts of the body to those beyond the ligature. This then is the difference between a tight ligature and a middling-tight one, that the tight ligature does not only intercept the passage of the blood in the veins but also in the arteries, whereas the middling-tight ligature does not hinder the pulsific force to such an extent that it cannot stretch beyond the ligature and drive the blood on into the furthest parts of the body.

We may therefore, advance the following argument. When in a middling-tight ligature we see the veins swollen and distended and the hand very full of blood, whence comes this? For either the blood comes through the veins or the arteries below the ligature, or through the hidden porosities of the flesh. From the veins it cannot come; through hidden porosities still less; therefore it needs must come through the arteries as I have said. That it cannot come through the veins is evident seeing that the blood cannot be squeezed back above the ligature unless you take the ligature quite away, when on a sudden all the veins are seen to subside and unburden themselves into the upper parts, the hand to grow white and all the previously gathered swelling and blood to vanish apace upwards.

He himself will perceive this better whose wrist or arm has been bound in this way for some time and whose hands in consequence have become swollen and somewhat colder, he indeed will feel, after the untying of the middling-tight ligature, something cold creeping up to his elbow or armpits, that is, along with the returning blood. This return of cold blood to the heart, when the ligature is untied after blood-letting, I have been inclined to think was the cause of fainting which we have seen happen even in strong men and chiefly at the moment of the untying of the ligature and which is commonly said to be occasioned by the 'turning' of the blood.

Moreover, when immediately upon the loosening of a tight ligature into a middling-tight one, we see that by the immission of blood through the arteries, the veins included under the ligature forthwith begin to swell and not the arteries, this is a sign that the blood flows from the arteries into the veins and not contrariwise, and that there is either an anastomosis of the vessels, or that there are porosities in the flesh and solid parts which are pervious to the blood. Again, a sign that very many veins communicate with each other is this, that when a middling-tight ligature is made above the elbow, many veins all together rise up and swell, but a passage for the blood being opened out of one little vein with a lancet, they all immediately subside and unburden themselves into that little one and almost all flap down straightaway.

From this every man can know the cause of the attraction which is produced by a ligature and perhaps also that of all fluxes, namely that the veins are compressed and the blood cannot get out, just as in the case of the hand as a result of the ligature I called middling tight. So that while the blood is being driven violently through the arteries, that is by the force of the heart, and not being able to get out, of necessity it must be that the part is filled with blood and distended.

For otherwise, how can it happen? Heat and pain and the force of a vacuum do indeed attract, but only so far that the part may be filled full and not that it should be distended and swollen beyond its natural constitution. But that the part should be so violently and so suddenly overwhelmed by reason of the in-breaking and hard and forcible in-driving of the blood that the flesh is seen to be torn into shreds and the vessels burst open, that this can ever be done either by heat or pain or the force of a vacuum is neither to be believed nor can it be demonstrated.

Moreover, it so happens that there is an attraction made by the ligature without either pain or heat or the force of a vacuum at all. Now if the blood should chance to be attracted by any pain, how could it happen when the arm is ligated at the elbow, that beyond the ligature the hand and the fingers swell and the veins become ingorged with blood, seeing that by reason of the pressure of the ligature the blood cannot reach them through the veins? And why should there not appear above the ligature any sign of swelling or repletion, or of engorgement of the veins or any trace at all of attraction or afflux of blood?

But the manifest cause of the attraction or afflux of blood beyond the ligature and of the preternatural swelling in the hand and fingers is this: that the blood enters forcibly and abundantly but cannot get out again. It may be asked whether this is not the cause of every tumour, as Avicenna says, and of every overburdening redundancy in any part. Because the ways of ingress are open and the ways of egress closed, the blood must needs become too abundant and make the parts swell up.

Again it may be asked whether from this does not also arise the fact that in inflammatory tumours, as long as the swelling is increasing and has not reached its final state, a full pulse can be felt in that place, especially in tumours that are hotter and in which the swelling usually occurs rapidly. But these are matters to be treated in a later discussion. Again, does this explain what I myself experienced by chance? Falling once from my carriage, I struck my forehead where a small branch of the artery creeps over from the temples, and immediately after the blow, in the space of about twenty pulses, I felt a swelling about the size of an egg which was neither hot nor very painful; clearly, on account of the nearness of the artery the blood was being driven more abundantly and more swiftly into the place I had bruised.

Hence indeed it does appear why in phlebotomy when we would have the blood leap out further and with greater force, we make the ligature above and not below the vein that is to be cut. For if the blood were to flow out from thence through the veins in so great abundance from the upper parts, a ligature of this kind would not help but hinder. And it were more reasonable to bind the ligature below so that the blood being checked might go out more profusely, if the blood did descend to this place from the upper parts through the veins and flow out from them. But since the blood is driven in different channels through the arteries into the lower veins through which its return is impeded by reason of the ligature, the veins swell and being distended can drive out the blood through the orifice with greater force and throw it further. But see, when the ligature is untied and the way open for its return, the blood no longer falls out but drop by drop. And as everybody knows, if in the process of blood-letting you untie the band, or bind it below, or constrict the limb with too tight a ligature, then the blood comes out without any force, because for sure, the way of entrance and influx of blood through the arteries is interrupted by that tight ligature, and when the ligature is untied, a freer passage is given it to return through the veins.

NOTES

1 Harvey is naturally still referring to the pulmonary arteries as arterial veins.

2 Cf. *De generatione animalium*, Ex. XIX: 'Bearing in mind this office of the arteries, the circulation of the blood, I have on occasion perfectly cured vast fleshy hernias beyond all hope by doing only this, namely, preventing the arrival at the affected part of any nutriment or spirit by cutting off or

ligating the arteriole, with the result that the fatal tumour was later easily extirpated by the knife or by fire. . . . A certain man . . . had in his scrotum a fleshy tumour or fleshy hernia larger than a man's head, hanging down to his knees. . . . Yet this vast excrescence . . . I completely removed by the method I have described and effected a perfect cure. . . .'

CHAPTER 12

That there is a circuit of the blood is shown from the proof of the second proposition

Seeing that these things are so, it is certain that another thing which I was saying before will also be proved, namely that the blood passes through the heart continually. For we see that the blood flows from the arteries into the veins and not from the veins into the arteries. We see, moreover, that almost the whole mass of blood can be drained from one arm (and that by opening but one cuticular vein with a lancet, provided the ligature be appropriately tied). We see also that it is poured out so forcibly and so abundantly, that not only the blood which before the incision was contained in the arm beyond the ligature is quickly and in a short time emptied out, but also the blood from the whole arm and the whole body, alike from the arteries and from the veins.

Therefore, it must needs be admitted first, that the blood is supplied with a strong driving force and that by force it is driven beyond the ligature, for it goes out with forceful vehemence and therefore this force derives from the strength of the pulse of the heart, for the force and impulsion of the blood comes only from the heart.

Second, it must needs likewise be admitted that this flow of blood comes from the heart, and that it arrives at this place having made a passage from the great veins through the heart, seeing that it enters the parts beyond the ligature through the arteries and not through the veins, and the arteries receive blood from the veins nowhere except out of the left ventricle of the heart. Otherwise, it were altogether impossible by any means whatsover to drain out so great a quantity of blood from one single vein, incised below the ligature, especially with such a rush, in such abundance, so easily and with such speed, unless it happened in consequence of the forceful power of the heart impelling the blood in the way I have said.

And if these things be so, then we may make a further calculation from them concerning the quantity of blood and bring forward a most clear argument touching the circular motion of the blood. For in venesection if anyone should suffer the blood to flow out for half an hour, the blood pouring forth with its usual impetuous rush, there is no doubt that when the greater part of the blood in the body was drained away, fainting and syncope would follow, and not only the arteries but the great veins also would be almost empty. It is therefore conformable with logic and reason to suppose that in the same space of half an hour, exactly the same amount of blood passes across from the great vein through the heart into the aorta. Furthermore, if you should reckon how many ounces of blood flow through one arm, or how many are thrust beyond a middling-tight ligature in twenty or thirty pulsations, it would assuredly allow you to estimate the whole amount: how much is passing during that same time through the other arm, through both legs and along both sides of the neck and through all the other arteries and veins of the body. To all these parts, the flow through the lungs and ventricles of the heart needs must incessantly bring up fresh blood, and as this perforce comes from the veins, the blood needs must make a circuit, seeing that it cannot be supplied by the food eaten and is more by far than is required for the nutrition of the parts.

It is to be observed further, that what sometimes happens during a venesection proves this truth. Although you may have ligated the arm correctly and made an incision with the scalpel in the proper way and opened the orifices and performed the whole operation aright, yet if an attack of fear or fainting from any other cause or disturbance of mind supervene and the heart beats more weakly, the blood will by no means issue forth except drop by drop, particularly if the ligature be made a little tighter. The reason for this is that the weaker pulse and feebler driving force has not strength enough to open the artery which is compressed and thrust the blood beyond the ligature. In fact, the heart, being weakened and enfeebled, cannot even drive the blood through the lungs or transfer it plentifully from the veins into the arteries. Thus in the same way and for the same reasons it happens that women's menses and haemorrhages of all kinds are checked. The same truth also appears from the reverse of these happenings, for when these persons come to themselves again, their presence of mind restored and their fear removed, and the pulsific strength of the heart increased, you shall immediately see the arteries beat more vehemently, even in the part where they are ligated, and move in the wrist, and the blood leap out further from the orifice in a continuous jet.

CHAPTER 13

The third proposition is proved and confirms the existence of the circulation of the blood

Thus far I have spoken of the quantity of blood which passes through the heart and lungs in the centre of the body and also from the arteries into the veins in the body at large. It remains for me to explain how the blood flows back from the extremities through the veins to the heart, and how the veins are the vessels carrying blood from the extremities to the centre only. That done, I think that the three propositions that I advanced as basic premisses for the circulation of the blood will be clearly and truly established and sufficient to win belief.

Now my third proposition will be made plain enough from the valves which are to be found in the hollow channels of the veins, and from their use, and from experiments demonstrated before your eyes.

The famous Hieronymus Fabricius ab Aquapendente,[1] a most learned anatomist, a man advanced in years and worthy of respect, or else Jacobus Silvius, as the learned Doctor Riolan would have it, first described these membranous valves in the veins as exceedingly thin, raised small portions of the inner coat of the veins and shaped in the figure of a sigma or half-moon. They are situated at diverse distances from one another, differently in different men, growing out from the sides of the veins and looking upwards towards the roots of the veins, and facing inwards towards the central hollow channel of the vein.

Both the valves, for they are for the most part in pairs, face each other and touch one another and so stick together with their tips that they are apt to be joined, so that if anything should slip through from the root of the veins into the branches, or from the larger branches into the smaller, they would altogether stop its passage. And they are so placed that the horns of the one that comes after look towards the middle of the curved part of the one above it, and thus they continue to alternate.

The discoverer of these valves did not rightly understand their use and other writers have added nothing. For their use is not to prevent the whole mass of the blood from falling downwards by its own weight into the lower parts of the body. There are indeed in the jugular veins valves looking downwards and preventing the blood from being carried upwards. They do not look upwards everywhere but always towards the roots of the veins and everywhere towards the region of the heart. I myself and others too, have on occasion found them in the renal veins and in the mesenteric branches looking respectively towards the vena cava and the portal vein.[2] Add to this also the fact that there are no valves in the arteries. It may be further noticed that dogs and all oxen have valves in the dividing of the crural veins at the beginning of the sacral bone, or in the branches near the hip-bone, and in these animals there is no such thing to be feared as that the blood should fall down by its own weight on account of their upright stance.

The valves are not set in the jugular veins for fear of apoplexy, as some allege, because in sleep any substance would be more apt to flow into the head through the carotid arteries.[3]

Nor are there valves so that the blood may stand still where the veins separate and in the tiny branches and not all rush into those that are more wide open and capacious, for they are placed where there are no divisions, though I admit that they are to be seen in greater numbers where divisions exist.

Their office is not to slow up the movement of the blood from the centre of the body. It is indeed more likely that it would go slowly enough of its own accord from the greater into the smaller branches and be separated from the whole mass and fountainhead,[4] or migrate from the warmer regions into the colder ones.

But valves were made entirely lest the blood should move from the greater veins into the lesser and so tear or swell them, that it should not go from the centre of the body to the extremities, but rather from the extremities to the centre. Therefore to this movement the delicate valves easily open a way, but the contrary movement they entirely arrest. And they are so positioned and arranged that if anything is not stopped in its passage by the horns of the upper ones, but slips as it were through a chink, it is caught in the convexity of the next one which is set transversely, and there it stays and can pass no further.

I have very often found in dissecting veins that if, beginning from the root of the vein, I put a probe towards the small branches, with all the skill I could, it could not be further driven by reason of the hindrance of the valves. But on the contrary, if I put it from the outer branches towards the root it passed very easily. In many places, a pair of valves, one on each side, is so sited and adjusted that when they are raised up, they close with one another in the middle of the hollow channel of the vein and join together so exactly, convexity to convexity and extremity to extremity, that you can neither see with your eyes nor in any way discover any crevice or conjunction. But on the contrary, they yield a passage to a probe inserted from the outside inwards, and like flood-gates which stay the course of rivers, are most easily pushed aside. These valves are so constituted that they hinder the movement of the blood outwards

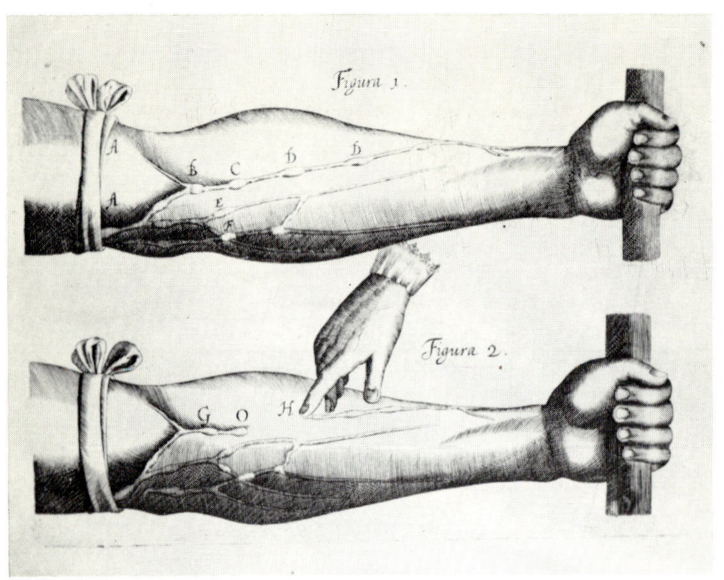

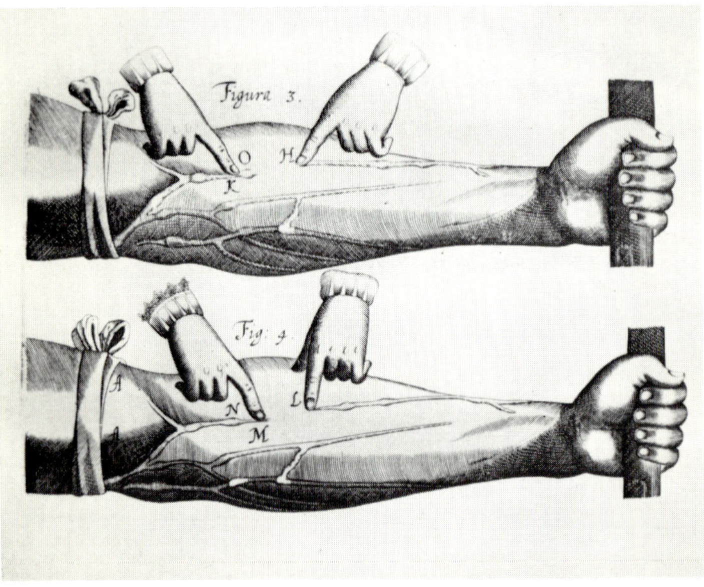

Plate 2. Harvey's experiments on the ligated arm

from the heart and the vena cava, and when in many places they are raised up towards each other and exactly closed, they altogether prevent and arrest it. Nor will they suffer the blood to move anywhere away from the heart, neither upwards to the head nor downwards to the feet, nor to the sides nor arms, but such indeed is their constitution that they stop and resist all movement of the blood which is begun in the greater veins and ends in the lesser. Yet any movement which begins in the small veins and ends in the greater, they obey and provide a free and open way for it.

But that this truth may the more clearly shine forth, let the arm of a living man be tied above the elbow (Fig. 1, AA) as if it were to let blood. There will appear at intervals, especially in country people and those who are swollen veined, certain little nodes or swellings (BCDDEF), not only where there is a branching (EF), but likewise where there is none (CD). These nodes are made by the valves. When they have appeared like this on the outside of the hand or forearm, if you press on the node with your finger or thumb (Fig.2, O) and draw the blood downwards from that node or valve to the next (Fig. 2, H), you will see that no blood can follow, the valve quite hindering it, and that the portion of the vein (Fig. 2, HO) between the node and the finger you moved down is quite obliterated and yet full enough above the node or valve (OG). Now, if you retain the blood so drove down and the vein emptied (H), and with the other hand press downwards in the direction of the valve (Fig. 3, K), the upper part of the vein being full (O), you will find that by no means can it be forced or driven beyond the valve (O); but the more you try to do this, the more you will see the vein swollen and distended at the node or valve (O), and yet below it the vessel remains empty (Fig. 3, HO).

Hence since anyone can try this experiment in many different places, it appears that the function of the valves in the veins is the same as that of the three sigmoid valves which are made in the orifice of the aorta and pulmonary artery, namely to shut so exactly that they do not allow the blood passing through to turn back again.

Furthermore, tying the arm again as before (AA), and the veins swelling, if you hold the vein at some distance below any node or valve (Fig. 4, L), and then with your finger (M) drive the blood upwards and above the valve (N), you will see that part of the vein (LN) remains empty and that the blood cannot return backwards through the valve (as in Fig. 2, HO). But taking away your finger (H), you will see it filled again from below (and be again as in Fig. 1, DC). From these things then it is certain and obvious that the blood moves in the veins upwards from the lower parts to those above, and towards the heart and not in the contrary direction. And although in some places there are valves which are not closed so precisely, or where there is but only one valve, and so do not seem to prevent entirely the flow of blood from the centre, yet for the most part it is evident that they do so, or at least, what seems negligently performed in one place appears to be compensated for either by the frequency or careful contrivance of the valves that follow in due order, or by some other way, so that the veins are open and patent ways for the blood returning to the heart but are entirely closed to blood going outwards from the heart.

In addition, this also is to be observed. Tying the arm as before and the veins swelling and the nodes or valves appearing, in a living man, if you press the vein firmly with your thumb at the spot below any one valve where you find the next, and hold it so that no blood can go from the hand upwards, then squeeze with your finger the blood from that part of the vein upwards to above the next valve (LN), as was said before; then taking away your finger (L), suffer the vein to be filled again from below (as DC), and then pressing again with your thumb in the same place, squeeze out the blood upwards (LN and HO), and do this a thousand times in a little space.

Now if you reckon the business, supposing how much blood in one compression has been moved upwards beyond the valve, and multiplying that by a thousand, you will find so much blood passed by this means through a little part of a vein in a short time, that I think you will find yourself perfectly convinced concerning the circulation of the blood and of its swift motion.

But lest you should say that by this experiment Nature is forcibly constrained, if you would do the same thing with valves set far apart and would observe, having taken away your thumb, how soon and how swiftly the blood runs upwards and fills the vein from below, I have no doubt that you will be convinced that it is so.

NOTES

1 Fabricius's account of the valves of the veins, *De venarum ostiolis*, was published from Padua in 1603. He claimed to have seen them for the first time in 1574 and from then onwards he demonstrated them to the medical students of Padua. He died in 1619 being then over 80.

Jacobus Silvius, or Jacques Dubois (1478–1555) had been one of the first professors to teach anatomy from dissection in Paris. Vesalius was among his pupils and although Silvius may not have been a consummate master in the art, Vesalius's abuse and ridicule of him would not seem from his writings to have been altogether justified.

It is generally believed on the authority of Salomon Alberti (*Tres orationes*, Nuremberg 1585) that the venous valves were pointed out to Vesalius in 1544 by Giambattista Canano of Ferrara. Vesalius, however, made nothing of this discovery, and does not describe the valves in the *Fabrica*.

2 One person to mention the existence of valves in the mesenteric veins was Realdus Columbus, *De re anatomica*, VI, p. 165. The purpose of these 'membranes' was in his opinion to prevent the chyle from returning to the intestines.

3 In the terminology of Harvey's time, the carotid arteries were still called the *soporales*, a name given to them from ancient times in the belief that they had an influence on sleep, presumably because pressure on them can produce a state of unconsciousness.

4 See above, p. 43 and below, p. 111.

CHAPTER 14

Conclusion of the demonstrative proof of the circulation of the blood

Now at last allow me to put forward my opinion concerning the circulation of the blood and to state it formally to all men.

Seeing that the following premisses have been proved by logical arguments and ocular demonstrations: that the blood passes through the lungs and heart by the pulse of the ventricles, and is driven in and sent into the whole body and there creeps into the veins and porosities of the flesh, and through the veins themselves returns from all parts, from the circumference to the centre, out of the tiny veins into the greater, and from thence comes into the vena cava and at last into the auricle of the heart, and in so great abundance, with so great an outflowing and inflowing, from hence through the arteries thither, from thence through the veins hither back again, that it cannot be furnished by those things that we eat, and in far greater abundance than is necessary for nourishment; it must of necessity be concluded that the blood is driven into a round by a circular motion in living creatures, and that it moves perpetually; and that this is the action or function of the heart, which by pulsation it performs; and lastly, that the motion and pulsation of the heart is the only cause.

CHAPTER 15

The circulation of the blood is proved by arguments from probability

It would not be irrelevant to the matter in hand to add these arguments also, because, according to certain processes of formal logic, a thing is so if it be both accordant with reason and necessary.

First (according to Aristotle, *De respiratione*[1] and *De partibus animalium*,[2] Bks II and III and elsewhere), since death is a corruption which befalls by reason of the defect of heat, and since all living things are warm and all dying things are cold, there must be a place for and a source of heat, a home and hearth as it were, wherein are contained and preserved the touchwood of Nature and the primordia of innate heat; from whence heat and life may flow as from their origin into all the parts and to which aliment may come and on which concoction and nutrition and all enlivening may depend.

That this place is the heart and that the heart is the foundation or first principle of life and that it is this in the way I have just said, I trust no one doubts.

There is, therefore, a motion required for the blood, and a motion of such a kind that it may return again to the heart, for having been sent into the outward parts of the body, far from its fountainhead (as Aristotle says, *De partibus animalium*, Bk II), and staying there motionless, it would congeal.[3] For we see that by motion heat and spirit are engendered and preserved in all things and by want of it they vanish. Seeing therefore that blood staying in the outward parts is congealed by the cold of the extremities and of the ambient air, and destitute of spirits, as it is in dead things, it was necessary that it should seek again both heat and spirit, and indeed its own preservation, from their fountainhead and origin, and by returning thither be once again made whole.

We see that by the exterior cold the extremities are sometimes chilled so that the nose and hands and cheeks look blue like those of dead men, and the blood in them takes on a leaden colour like that blood which usually settles down in the pendent limbs of corpses, and the limbs become so numbed and hardly moveable that they almost seem to have lost their life. They could certainly by no means recover again their heat and colour and life, especially so soon, unless they were warmed by a new in-flowing and in-driving of heat from the source of heat. For how can they attract when heat and life are almost extinct in them? Or, when their passages are narrowed down and filled with congealed blood, how can they let in the incoming aliment and blood, unless they first discharge their contents? and unless the heart were indeed the first principle of these things, where life and heat remain though the parts are chilled (as Aristotle says, *De respiratione*, cap. 2), and unless the blood which has become cold and effete be driven forth by new blood, imbued with heat and spirits, coming from the heart and sent through the arteries, that all the parts may renew again their languishing heat and vivifying warmth that almost was extinct?

Thus it is that while the heart itself remains unharmed, life can chance to be restored to all the parts and health recovered. But if the heart be either chilled or affected with some grievous ill, it needs must be that the whole animal will suffer and fall into corruption. When the first principle is corrupted and impaired, there is nothing (as Aristotle says, *De partibus animalium*, Bk III),[4] which can afford help either to it or to those other things which depend upon it. And from this perhaps comes the reason why in those that are oppressed with grief, or love, or envy or other cares of this kind, there befalls a consumption and a wasting away, or a cacochymia and cachexia which cause all kinds of diseases and kill men. For every passion of the mind which troubles men's spirits either with grief, joy, hope or anxiety, and gets access to the heart, there makes it change from its natural constitution in temperature, pulsation and the rest,[5] and so pollutes the whole aliment in its source and impairs its strength. It should be no means seem a strange thing that thereupon it begets diverse sorts of incurable diseases in the limbs and in the body, seeing that the whole body in that case is afflicted by the corruption of the aliment and the deficiency of innate heat.

Moreover, as all animals live by aliment inwardly concocted, it is necessary that the concoction and the distribution be perfectly accomplished and therefore there must be present a place and a receptacle where the aliment is perfected and from whence it is distributed to each member of the body. Now this place is the heart, for it alone of all the parts contains blood not only for its private use in the coronary vein and artery, but also for the public or general use of the body in its cavities, that is in its auricles and ventricles, as in cisterns or a storehouse, but the rest of the parts have it in vessels only for their own advantage and private use,[6] and also because the heart alone is so placed and appointed that from thence by its pulse it may dispense and distribute the blood equally to all the parts, and that according to the just proportion of the size of the channels of the arteries which supply each part, and in this way may give to those that want as out of a treasurehouse and fountain.

Furthermore, for this distribution and motion of the blood, force and vehemence are required and an instigator of this force, such as the heart is. This is either because the blood of its own accord easily concentrates, that is runs towards a common centre as if towards its beginning,[7] or like a part to the whole, and coalesces as drops of water spilt upon a table run together into a mass, and this it is wont to do very quickly and for slight causes such as cold, fear, shivering and the like; or else it is because the blood, seeing that it is squeezed out of the hair-like veins into the little branches and from thence into the greater by the movement of compression of the limbs and muscles, is more liable and more prone to move from the circumference to the centre than in the opposite direction, even if there were no valves to prevent the contrary movement. From which it follows that in order to leave its beginning and enter into places that are narrower and colder, and be moved against its natural inclination, the blood has need both of vehemence and a driving force such as is the heart alone, and that in the way which has already been said.

NOTES

1 The reference is to that part of the *Parva naturalia* now called *De iuventute et senectute* and which corresponds to the earlier chapters of *De respiratione*, according to the older divisions of Aristotle's works. See chapter, 4, 469b 5–20: 'But in animals all the members and the whole body possess some connate warmth of constitution, and hence when alive they are observed to be warm, but when dead and deprived of life they are the opposite. Indeed, the source of this warmth must be in the heart in sanguineous animals . . ., for, though all parts of the body by means of their natural heat elaborate and concoct nutriment, the governing organ takes the chief share in this process. Hence, though the other members become cold, life remains; but when the warmth here is quenched, death always ensues, because the source of heat in all other members depends on this. . . . Hence, of necessity, life must be coincident with the maintenance of heat, and what we call death is its destruction.'

2 As Harvey says there are a great many references to these opinions in *De*

partibus animalium. Cf. 650a 5: 'since everything that grows must take nourishment and since by the force of heat this is concocted, it follows that all living things must have a natural source of heat'; 670a 25: 'there must be some part which like a hearth, shall hold the kindling fire, and this place is the heart'. Harvey had already expressed these views in his *Anatomical Lectures*, see pp. 244, 251, and in his *De motu locali animalium* had said, p. 103: 'A failing of the spirit results in solidification, congealing (it brings torpor), lifelessness and death.'

3 *P.A.*, II, 3, 649b 25: 'Blood in a certain sense is essentially hot . . . for heat is included in the definition of blood. . . . But so far as blood becomes hot from some external influence, it is not hot essentially. . . . Some substances are hot and fluid so long as they remain in the living body but become perceptibly cold and coagulate as soon as they are separated from it.' 650b 2: 'the purpose of the blood in sanguineous animals is to subserve the nutrition of the body'.

4 *P.A.* III, 4, 667a 35: 'if the primary part be diseased there is nothing from which the other parts which depend upon it can derive succour'.

5 Though Harvey adopts these ideas as his own, they are mostly to be found in Aristotle. Cf. his *De motu animalium*, cap. 8.

6 Cf. Aristotle, *P.A.* 66a 1–5.

7 Cf. *De motu locali animalium*, p. 103: 'Just as the senses move the appetite so the appetite sets in motion the spirit. But the appetite arises from the heart and returns to the heart, for it exists entirely in the emotions, anger, fear, etc., and every emotional state is accompanied by concentration or expansion or ebullition. Moreover all concentration is towards the heart. Because the heart is the centre.'

CHAPTER 16

The circulation of the blood is proved ex consequentiis

As consequential upon this truth which has been postulated there are various unsolved problems which taken as arguments *a posteriori* are of some value in confirming its validity. Though these questions and others seem to be enveloped in much doubt and obscurity, yet they can easily be assigned causes and a logical explanation from my hypothesis.

How does it happen that the whole body is vitiated when the small part that is infected remains unharmed? as we see in contagion, in a poisoned wound and in the bite of a snake or of a mad dog, in the pox and the like. The pox often betrays itself first of all by pain in the shoulders or in the head or by other symptoms, while the genitalia remain unaffected. And the wound made by the bite of a mad dog being healed, yet I have seen fever and the other terrifying symptoms ensue. The reason is because the contagion at first imprinted on the part is carried along with the returning blood to the heart, and hence it is clear that it can afterwards infect the whole body. In the beginning of a tertian fever, the morbific cause seeking the heart is delayed around the heart and lungs, and makes the sufferer short of breath, panting and spiritless, because the vital principle is oppressed and the blood is stuck in the lungs and congealed and does not pass through (I say this as one who has had experience in the dissection of those who have died at the beginning of an attack), at which times the pulse is always small and fast and sometimes irregular. But when the heat is increased and the matter is thinned in consistency, the ways opened and the transit of the blood accomplished, the whole body grows hot, the pulses become greater and more vehement and the paroxysm of the fever occurs, that is to say, when the preternatural heat is kindled in the heart, it is diffused from thence through the arteries into the whole body together with the morbific matter which by this means is overcome and dissolved by Nature.

Again, why is it that medicaments outwardly applied exert their powers inwardly, just as if they had been taken by mouth? For example, colocynth[1] and aloes loosen the belly; cantharides moves the urine; garlic applied to the soles of the feet causes expectoration; cordials strengthen the heart, and there is an infinite number of others of this kind. It is to be supposed that the veins absorb through their orifices some little of those things which have been applied outwardly and carry it in with the blood, after the same way as the mesenteric veins suck chyle from the intestines and carry it to the liver along with the blood, and to say this is perhaps not contrary to reason and logic.

In the mesentery also, the blood entering through the coeliac artery and through the superior and inferior mesenteric arteries goes on to the intestines, from which together with the chyle which has been attracted into the veins, it returns through their very many ramifications to the porta hepatis and through it into the vena cava. And so it happens that the blood in these veins is imbued with the same colour and consistency as in the rest, contrary to what a great many believe. Nor is it necessary to believe that two contrary motions go on in each hair-like branch of the veins, one of chyle upwards, the other of blood downwards, a thing which is neither accordant with reason nor probability.[2] But is it not rather that Nature with her consummate foresight acts in this way? If the chyle were mixed with the blood, the raw chyle with the concocted blood in equal proportion, no concoction, transmutation or sanguification would arise from thence, but rather, since they are active and passive in relation to each other, there would result from the union of both of them a mixture and a thing of middle nature between them, as in the mingling of wine with water when sour wine results. But now, when with the great quantity of blood that passes by, a tiny portion of chyle is mixed in this way and as it were in no noticeable quantity, the same thing happens of which Aristotle speaks saying that when one drop of water is added to a hogshead of wine, or contrariwise, the whole is not a mixture but either wine or water. So when the mesenteric veins are dissected, there is not found in them either chyme, or chyle and blood, either separately or mixed together, but blood that is visibly of the same colour and consistency as that in the other veins. But because there is nevertheless a small something of unconcocted chyle, although imperceptible, Nature added the liver, in whose meandering channels the blood is delayed and receives a fuller transmutation, lest coming prematurely to the heart still raw, it should overwhelm the principle of life. Hence, in an embryo, there is scarce any use of the liver and so the umbilical vein clearly passes right through it without branching, and from the porta hepatis there exists a foramen or anastomosis so that the blood returning from the intestines of the foetus goes to the heart, not through the liver, but running through the aforesaid umbilical vein together with the maternal blood returning from the placenta of the uterus.[3] Hence it happens also that in the first formation of the foetus the liver is formed later, and I have myself observed in a human foetus all the limbs perfectly delineated, even the genitalia distinct, when as yet the rudiments of the liver had scarcely been laid down.[4] And truly, as long as the limbs appear white, as even the heart itself does in the beginning, and there is no redness at all contained except in the veins, you will see nothing in the place of the liver except a shapeless mass as of extravasated blood which you might think to be some bruise or broken vein.

There are in an egg as it were two umbilical vessels, one from the white which passes unbranched through the liver and runs directly to the heart, the other which goes from the yolk and ends in the portal vein. For it happens in an egg that the chick is first formed and nourished from the white and by the yolk after the chick has been perfected and even hatched, for the yolk can be found contained among the intestines in the belly of the chick for many days after it has been hatched, and the yolk corresponds to the nourishing milk of other animals. But I shall discuss these things more appropriately in my observations concerning the formation of the foetus where there may be many problems of this kind, such as why this is made or perfected first and that later, and in regard to the hierarchy of the members, which part is the cause of another.[5] And concerning the heart there are many problems as why, as Aristotle says in *De partibus animalium*, Bk III, the heart was made and established first and why it seems to have life in it and motion and sensation before any other part of the body is perfected. And concerning the blood also, as why it exists before everything else and in what manner it has in it the beginning of life and of the animal and requires to be moved and driven up and down for which reason the heart seems to have been made.

In the same way, in the observation of pulses, we may enquire why some indicate death and others the contrary, and by considering the causes and presages in all kinds of pulses, to find out what these signify, what those, and why. And we may enquire into all crises and natural purgations, into nutrition, especially into the distribution of nutriment, and also into all fluxes and so forth.

Lastly, in all parts of physic, physiological, pathological, semeiotic, therapeutic, when I consider with myself how many questions may be answered, given this light and truth, how many doubts may be resolved, how many obscure things made clear, I find a most large field where I might run out so far and enlarge so widely that not only would this work grow into a great volume, which is not my intention, but perhaps even my lifetime would not suffice to make an end of it.

Therefore in this place, that is to say in the following chapter, I shall endeavour to refer to their proper uses and true causes, only those things relating to the structure of the heart and arteries which are visible in the course of an anatomy, seeing that there will be found there, as in whatsoever direction I might turn my attention, very many things which receive light from this truth and in return make this truth to shine more bright. And so it is my wish that above all else it be established and supplied with arguments based on anatomy.

There is one thing which it will not be irrelevant to take notice of here in passing, although it ought to have a place among my observations on the use of the spleen.[6] From the upper part of the branch of the splenic vein that runs in the pancreas arise the posterior coronary, gastric and gastro-epiploic veins, all of which with their many branches and ramifications are distributed in the stomach as the mesenteric veins are to the intestines. In the same way, from the lower part of this branch of the splenic, the haemorrhoidal vein runs down to the colon and rectum. The blood returning through these veins on both sides carries with it from the stomach on the one hand a juice that is somewhat raw, watery and thin, the process of chylifaction having not as yet been perfectly completed, and on the other from the guts a juice that is thick and more earthy as coming from the faeces. In this branch of the splenic vein, by the mixture of contraries, they are suitably adjusted to one another, and Nature by mixing together both these juices, each of more difficult concoction because their natural dispositions are contrary, and having poured on them a very great quantity of warm blood which by reason of the multitude of the arteries flows most copiously from the spleen, brings them all now better prepared to the porta hepatis and by means of such a structure of the veins supplies and compensates the deficiency of both extremes.

NOTES

1 Colocynth is a plant of the gourd family, otherwise known as the Bitter-apple. It was used either in the form of a pulp or a resin and is a violent purgative. Cantharides, the Spanish Fly, when used as a vesicant gives rise to urinary symptoms. It was also used as an aphrodisiac, but heavy doses taken internally produce haematuria.

2 The traditional Galenic view was that food taken into the stmoach was there concocted and turned into chyle which entered the mesenteric veins and the portal vein and so passed to the liver where it was further concocted into blood. At the same time the stomach received a supply of venous blood for its nourishment from the liver through the same portal vein. Strictly speaking, chyle is absorbed by the guts from chyme which is the product of digestion to be found in the lumen of the intestines.

3 See the section on the uterine membranes and humours in Harvey's *De generatione*. He considered that the placenta served the office of the liver for the foetus. The 'foramen or anastomosis' is the ductus venosus through which blood, carried from the placenta by the umbilical vein, passes into the inferior vena cava without going through the liver.

4 *Anat. Lect.*, p. 127: 'I have seen . . . a body with all parts perfected yet with a liver unformed; a heart fashioned with auricles but the liver ill-made and a shapeless mass.'

5 All these problems are discussed at length in *De generatione*.

6 Harvey had already discussed this problem in his *Anatomical Lectures* and reached a different conclusion. The question is the respective roles of the liver and the spleen in the process of sanguification: 'Because animals like man eat food made up of different parts of which some are of easy concoction and other of more difficult, . . . for the former the liver is necessary; but that part which runs back to the portal veins . . . is diverted and attracted to the spleen where it can be received, concocted and perfected by means of the abundant heat and loose texture of the part. And this is seen to be proved by the structure of the veins . . .' (p. 129).

In this whole passage Harvey is guilty of arguing from suppositions which he cannot prove and which are untrue. However, the existence of the portal circulation follows by inference from the general circulation. His proof of it will be found in his Second Letter to Riolan (pp. 192–3).

CHAPTER 17

The hypothesis of the movement and circulation of the blood is proved by those things which are to be observed in the heart and by those which are to be seen in anatomical dissection

I do not find the heart in all creatures to be a distinct and separate part, for some plant-animals, as you might say, have no heart. Because some animals are colder, with a very tiny body, soft in texture, of a kind of homogeneous composition like the race of caterpillars and worms and those very many creatures that are born of putrefaction and do not keep their same outward appearance,[1] in these there is no heart as being creatures that have no need of a driving force to dispatch the nutriment to the extremities. For they have a body which grows all at one time and is an undivided entity without limbs, so that by the contraction and relaxation of the whole body they take in food and expel it, move it and remove it. Plant-animals so called are oysters, mussels, sponges and the whole race of zoophytes and they have no heart for they use their whole body in the place of a heart and an animal of this kind is as it were entirely a heart.

In very many or nearly all kinds of insects we cannot see the heart properly on account of the exceeding smallness of the body. However, in bees, flies, hornets and the like, it is possible sometimes with the help of a magnifying glass, to make out some pulsating thing. And in lice also, in which with the help of a magnifying glass, you may in addition clearly watch the passing of food through the guts like a black spot seeing that the creature is transparent.[2] But in some bloodless and colder animals like snails, molluscs, shrimps and crustaceans, in all these there is a tiny pulsating part, like a kind of bladder or an auricle without a ventricle, but which makes its contraction and beat rather infrequently and which can only be discerned in summer or in warmer weather.

In these creatures this small part is made in this way because they need some kind of driving force to distribute the food on account of the variety of the organic parts or the density of their substance. But the pulsations are made rather infrequently and sometimes not at all, by reason of their coldness, as is appropriate to them seeing that they are of an ambiguous nature so that they seem sometimes to live, sometimes to die, and sometimes to live the life of an animal and sometimes that of a plant. This is also seen to happen in insects which lie hidden and concealed in winter as if they were dead, or were leading only the life of a plant. But whether it also happens to some sanguineous creatures, like frogs, tortoises, snakes and swallows,[3] we may not unreasonably take leave to doubt.

In animals which are bigger and warmer, because they are sanguineous there is need of something to impel the food, something that is perhaps endowed with greater force. Therefore, in fish, snakes, lizards, tortoises, frogs and the like, the heart has both one auricle and one ventricle from which follows that most true statement of Aristotle in *De partibus animalium*, Bk III,[4] that no creature having blood lacks a heart, and therefore by this stronger and more robust impelling force, the aliment is not only put in motion by the auricle, but thrust our further and more swiftly.

In animals that are still larger, warmer and more perfected, as abounding in much blood that is hotter and full of spirit, a stronger and more fleshy heart is required to thrust forth the aliment more powerfully, more swiftly and with greater force by reason of the magnitude of the body or the density of its texture.

And furthermore, because the more perfected creatures need a more perfected aliment and a more abundant innate heat, and so that their aliment may be concocted and acquire a further perfection, it was fitting and reasonable that these animals should have lungs and a second ventricle which should drive the aliment through the lungs themselves.

So in whatsoever creatures there are lungs, the heart has two ventricles a right and a left, and wheresoever there is a right ventricle there is also a left and not the reverse that where there is a left there is also a right. I call the left ventricle that which is distinct by its use and not by its position, namely, the ventricle which pours out the blood into the whole body and not only into the lungs. Hence the left ventricle by itself seems to make up the heart. And it is situated in the midst of the heart, so carved out with deeper furrows and fashioned with greater care that the whole heart seems to have been made for the sake of the left ventricle, and the right ventricle seems as it were a servant to the left. The right ventricle does not reach to the apex of the left, and it has a wall that is three times as thin and has as it were a kind of articulation, as Aristotle says, on the top of the left. It is indeed more capacious for it not only provides material for the left ventricle but also aliment for the lungs.

It must, however, be observed that in an embryo these things are far otherwise and there is no such great difference between the ventricles, but like twin kernels in a nut, they are almost equal and the apex of the right reaches to the top of the left so that in them the heart has as it were a double apex to the cone.[5] And matters are arranged thus in embryos because, as I have said, while the blood does not pass through the lungs, and as it still goes from the right ventricle of the heart into the left, both ventricles alike perform the same office of bringing the blood across from the vena cava into the great artery by means of the foramen ovale and the ductus arteriosus, as I have already explained, and both alike drive it into the whole of the body and from this comes their similar constitution. But when the time comes for the lungs to be used and the aforesaid passages between the vessels to be stopped up, then the difference between the ventricles in strength and the rest begins to occur, because the right drives the blood through the lungs only and the left through the whole body.

In addition, there are in the heart small lizard-shaped muscles, if I may so call them, or little twigs of flesh and very many fibrous connections, which Aristotle in his book *De respiratione* and in *De partibus animalium*, Bk III, calls sinews,[6] some of which are stretched out separately in different ways and some are hidden in the walls and septum of the heart where there are deep grooves in which they lie as in a furrow, like a kind of small muscle. They are added to the heart to supplement, as it were, its contraction and so to give a stronger impulsion to the blood and to assist the heart in driving out the blood to a greater distance and, like the careful and ingenious provision of ropes in a ship, to help the heart from all sides when it contracts itself down in every direction and to expel the blood more fully and more forcibly out of the ventricles.

Now this is indeed clear from the fact that they exist in some animals and in some not, and in all animals which have them they are more numerous and stronger in the left ventricle than in the right. In some animals they exist in the left ventricle and not at all in the right. In men there are more in the left ventricle than in the right and more in the ventricles than in the auricles; in some there are practically none in the auricles. In the brawny and muscular bodies of country-folk and in those of a harder bodily build, they are numerous, but fewer in the soft bodies of women.

In those creatures in which the ventricles are smooth inside and altogether without fibrous bands and not cut into grooves, as in almost all little birds, snakes, frogs, tortoises and the like, and also in the partridge and the hen and likewise for the most part in fish, in these there are not found in the ventricles either those tendinous strips or so-called fibres, or the tricuspid valves. In some creatures the right ventricle is smooth inside, but the left has those fibrous bands, as in the goose, the swan and larger and heavier birds.

The reason for these differences is the same as in all other creatures: seeing that their lungs are spongy and loosely-textured and soft, they do not need so great a force to impel the blood through them, and therefore in the right ventricle either there are none of those fibres or they are fewer and weaker and not so fleshy or comparable to muscles. But in the left they are stronger and more in number, more fleshy and more muscular, because the left ventricle has need of more strength and force with which it must pursue the blood through the whole body.

For this reason also the left ventricle occupies the middle of the heart and its walls are three times thicker and stronger than those of the right. Hence for all animals, and for man also, the thicker, harder and more solid they are in fleshly build, and the more fleshy and muscular their hands and feet and the greater their distance from the heart, so are their hearts more fibrous, thick, robust and muscular. And this is an obvious necessity. On the other hand, the more finely-textured they are, the softer in build and the slenderer, so do they have a heart that is more flaccid and softer, less fibrous inside or not at all fibrous, and weak.

Consider likewise the use of the sigmoid valves which are so made as to prevent the blood once sent forth from returning again into the ventricles of the heart. They are situated in the orifice of the pulmonary artery and in the aorta, and when they are raised up, they meet together in a three-cornered line such as is left by the bite of a leech, and so the more tightly they are pressed together the more they prevent the return of the blood. The tricuspid valves in the entrance from the vena cava and from the pulmonary artery[7] are the door-keepers to prevent the blood from slipping back again when it has been driven on with the greatest possible force. For that reason they are not present in all animals, as I have said, nor, in those animals in which they are present, do they seem to have been made by Nature with the same skill; but in some they are made to close more exactly and in others more loosely and carelessly according to the greater or lesser driving force arising from the contraction of the ventricles. In the left ventricle therefore, so that there may be a more exact closure to withstand the greater driving force, there are only two valves shaped like a mitre and so that they may shut most precisely they stretch for a long way through the midst of the ventricle into the apex. This it was perhaps which deceived Aristotle and made him think that this ventricle was double when he had cut it in a tranverse section.[8] It is likewise certain that it was to prevent the blood from slipping backwards into the pulmonary vein and thereby impairing the strength of the left ventricle in driving on the blood into the whole of the body, that these mitral valves much surpass in size and strength and precision of closing those which are placed in the right ventricle.

Hence, likewise of necessity, there is to be seen no heart without a ventricle since it must serve as cellar, source and storehouse for the blood. But the same does not always happen in the brain for almost all kinds of birds have no ventricle in the brain as is clear in the goose and the swan. Although their brains are almost equal in size to that of a coney, yet conies have ventricles in their brains, but the goose has none.[9]

Likewise, again, wherever the heart has one ventricle, there hangs beside it one auricle, flaccid, skinny, hollow within and full of blood. Where there are two ventricles, there are likewise two auricles. On the other hand, in some animals there is only an auricle and no ventricle to the heart, or at least a bladder that is analogous to an auricle, or the vein itself being dilated in that place makes the pulse, as can be seen in hornets and bees and other insects. I believe I can demonstrate by some experiments, that these have not only a pulse but a respiration likewise in that place which they call the tail, and this explains why this tail is lengthened and contracted sometimes more frequently and sometimes more seldom according as they seem to be more panting and in greater need of air. But I will speak of these things in my treatise on respiration.[10]

It is also clear that the auricles pulsate, contract themselves and, as I said before, cast the blood out into the ventricles, so that wheresoever there is a ventricle, there needs must also be an auricle not only, as is commonly believed, to be the receptacle and storehouse for the blood (for what need is there for pulsation for the retaining of blood?), but because the auricles are the first movers of the blood, especially the right auricle, being, as I said before, the first thing that lives and the last that dies, and therefore the auricle is necessary as being that which serves to pour the blood into the ventricle. And the ventricle immediately contracting itself, squirts out more completely and drives on more violently the blood that is already in motion, just as when you play at ball, you can strike the ball further and more strongly if you take it *à la volée* than if you simply throw it. And this again is contrary to the common opinion, because neither the heart nor anything else can so distend itself as to draw anything whatsoever into itself in its diastole, except while returning to its former constitution if it has been previously compressed like a sponge. But it is certain that all local movement in animals comes first and takes its beginning from the contraction of some particle.[11] Therefore, by the contraction of the auricles the blood is cast into the ventricles, as I have already made clear, and from thence by the contraction of the ventricles it is thrown further and driven into the arteries.

This truth concerning local movement, and that the immediate motive organ in every movement of all animals in which there is from the beginning a motive spirit, as Aristotle says in his book *De spiritu* and elsewhere, is contractile, and in what way *neuron* is derived from *neuo*, that is I nod, I contract, and that Aristotle did recognize muscles and not incorrectly referred every movement in animals to the nerves or to that which is contractile and therefore called those muscular bands in the heart nerves, all this I think will be made clear if at any time I shall have liberty to demonstrate from my own observations these matters concerning the motive organs of animals and the structure of the muscles.[12]

But to pursue my purpose concerning the use of the auricles which, as I have previously demonstrated, is to fill the ventricles with blood. It happens that the denser and more compact the ventricular part of the heart and the thicker its walls, the more resilient and the more muscular are its auricles to drive in the blood and fill the ventricles. On the other hand, in those creatures in which the auricle is as it were a bladder of blood or a membrane containing blood, as it is in fish, in them the bladder which is in lieu of the auricle is exceedingly thin and so large that the heart seems to rest upon its surface. And in those fish in which this bladder is a little more fleshy, it seems very prettily to emulate and counterfeit the lungs, as in the carp, the barbel and the tench and others.

In some men, that is in such as are brawny and of a harder build, I have found the right auricle so strong and so neatly made up within with fibrous bands and various interweaving of fibres, that it seemed to be the equal in strength of the ventricles of other men, and truly I was amazed that in different men there should be so great a difference.

But it is to be observed that in the foetus, the auricles are much larger in proportion because they are present before the rest of the heart is made or assumes its own function, as I have already shown, and there they perform as it were the office of the whole heart.

But the things which I have observed in the formation of the foetus and which I mentioned before and which Aristotle confirms in an egg, afford the greatest possible light and proof in this business. At first, when the foetus is still as it were a soft worm and while it is, as they say, in the milk, there is in it only a speck of blood, or a little bladder which beats and a portion of the umbilical vein, as it were, dilated where it begins or at its base. Afterwards, when the foetus is outlined and begins to have a harder bodily form, this little bladder becomes more fleshy and robust, and changing its constitution turns into the auricles above which the body of the heart begins to grow, though as yet it performs no public office. But when the foetus is completely formed and the bones are become distinct from the flesh and the animal is perfected and is seen to have movement, then also the heart is to be found beating within it, and as I have already said, passing the blood through both the ventricles from the vena cava into the great artery.

So Nature being perfect and divine and making nothing in vain, neither gave a heart to any animal where there was no need, nor made it before it could be of any use, but, by the same gradual stages in the formation of each animal and passing through the same structural constitutions, so to speak, common to all animals, that is egg, worm and foetus, it acquires perfection in each one. These things concerning the formation of the foetus will be proved elsewhere by a great many observations.

Now Hippocrates in his book *De corde*, did not without reason call the heart a muscle seeing that the action and function of both is the same, that is to contract itself, to move something else, namely the blood that it contains. Moreover, we may see the action and use of the heart from the composition of its fibres and its structure designed for movement just as in the muscles themselves. All anatomists have observed with Galen that the body of the heart is constructed of various sets of fibres, that is, straight, transverse and oblique, but in a heart that has been thoroughly boiled, the structure of the fibres is found to be quite different. For all the fibres in the walls and septum are circular as they are in a sphincter, but those which are in the muscular bands are stretched out at length obliquely. And so it happens that when all the fibres are contracted at the same time, the tip needs must be drawn up towards the base by the muscular bands, the walls are closed in on all sides into a round and the heart is contracted down from every direction and the ventricles are narrowed, and therefore, since its action is contraction, we must needs judge that its function is to thrust the blood out into the arteries.

And we must not disagree with Aristotle concerning the principality of the heart by asking whether it receives motion and sensation from the brain, or blood from the liver, or whether it is the beginning of the veins and of the blood and so forth. For those who endeavour to confute him in this way omit the chief argument, or fail to understand it, namely that the heart is in existence from the beginning and contains within itself blood, life, sensation and motion before either the brain or the liver was made or clearly appeared as a separate part, or at least before they could perform any function. The heart with its proper organs fashioned for movement, like some internal animal, exists from an earlier time. And having made it first, Nature would have the whole animal thereafter to be made, nourished, preserved and perfected as the work and habitation of the heart. And the heart like the Prince in the Commonwealth in whose person lies the first and supreme power, governs all things everywhere, and from it as from its origin and foundation in the living creature all power derives and on it does depend.[13]

There are, moreover, many things also concerning the arteries which shed light more fully upon this truth and prove it, as for instance, why the pulmonary vein does not beat although it is counted among the arteries, or why there is a pulse in the pulmonary artery, seeing that the pulse in the arteries arises from the impulsion of blood. The reason why the arteries differ so much from the veins in the thickness and strength of their coats is because they withstand the force of the heart as it impels the blood, and of the blood as it rushes forth. Hence, since Nature who is perfect makes nothing in vain and in all her works suffices every need, the nearer the arteries are to the heart, the more they differ from veins in their constitution and are more robust and ligamentous; but in their ultimate ramifications, in the hand, foot, brain, mesentery and spermatic vessels, they are so similar in their constitution that by looking closely at their coats it is a hard business to know one from the other. But this is so for very good reasons for the further the arteries are from the heart, with so much the less force are they struck by the stroke of the heart which is weakened by the great distance. Add to this that the driving force of the heart, since it must needs be sufficient to send the blood into all the trunks and branches of the arteries, is lessened at every partition as being divided up, insomuch that the final hair-like divisions of the arteries look like veins not only in their constitution but also in their function, seeing that they do not have a perceptible pulsation either at all or only sometimes, and only when the heart is beating more vehemently or some little artery is dilated or more open in some particular way. Hence it comes that sometimes we can feel a pulse in the teeth and in tumours and sometimes in the fingers, and sometimes we cannot. And so by this token I have been able to observe that children whose pulses are always swift and frequent, were in an undoubted fever, and likewise in tender and delicate people by gripping their fingers I could easily perceive by the pulse of their fingers when the fever was at its height.

On the other hand, when the heart beats more languidly, it is impossible to feel the pulse not only in the fingers but also in the wrist and in the temples, as in fainting, in hysterical manifestations, in asphyxia, in the more sickly and in those about to die.

And surgeons should be warned of this in case they are deceived, because in the amputation of limbs, in the excision of fleshy tumours and in wounds, the blood does indeed always flow with force from the artery, but it does not always come in spurts because the tiny arteries do not beat, especially if they have been compressed by a ligature.

Furthermore, it is for the same reason that the pulmonary artery not only has the constitution and coat of an artery, but also that it does not differ so much in the thickness of its coat from the veins as from the aorta. The reason is that the aorta withstands a greater impulsion of blood from the left ventricle than the pulmonary artery does from the right, and in the constitution of its coat it is as much softer than the aorta as the right ventricle of the heart is weaker than the left in its walls and fleshiness. And again, by as much as the lungs depart in the softness of their texture from the general build of the body and of its fleshiness, by so much does the coat of the pulmonary artery differ from that of the aorta. All these things keep always and everywhere a like proportion. In men, the more brawny and muscular and harder they are in build, and the stronger, thicker, more dense and more fibrous their heart, so do their auricles and arteries correspond in thickness and strength in due proportion with all the rest. Hence, in those creatures, the ventricles of whose hearts are smooth within, without villi or valves and with a thinner wall, as in fish, birds, snakes and a great many other kinds of animals, in them the arteries differ little or nothing from the veins in the thickness of their coats.

And furthermore, the reason why the lungs have such large vessels, the pulmonary artery and the pulmonary vein, so that the trunk of the pulmonary vein exceeds both the branches of the crural and jugular veins, and why the lungs are filled so full of blood as we know from experience and our own eyesight, not having been deceived by the inspection of lungs removed from dissected animals whose blood had completely drained away (a thing of which Aristotle gives warning), the reason is because in the lungs and in the heart is the storehouse, fountain-head and treasury of the blood and the laboratory of its perfection.

Likewise, the reason why we see in an anatomical dissection that the pulmonary vein and the left ventricle abound with so great a quantity of blood and indeed of the same blood, alike blackish and lumpy, as that with which the right ventricle and pulmonary artery are filled, is because the blood passes from one side to the other continually through the lungs.

Lastly, the reason why the pulmonary artery is commonly said to have the constitution of an artery and the pulmonary vein that of a vein, is because in very truth the former is an artery and the latter a vein in function, constitution and all things else, contrary to what is commonly believed. And why the pulmonary artery has so wide an orifice is because it carries so much more blood than were necessary for the nourishing of the lungs.

All these phenomena which can be observed during dissection and a great many more, if they be rightly weighed, are seen to shed light abundantly on the truth that I have stated and to prove it completely, and at the same time to go against the commonly received opinion, for it were very hard for anyone to explain by any other way than I have done for what cause all these things were so made and appointed.

NOTES

1 According to Aristotle creatures born of putrefaction arose spontaneously and in the shape of a worm which in its turn gave rise to an insect. In his *De generatione*, Ex. LXII, Harvey denies this and maintains that all living creatures without exception arise from a primordium that is itself living, that is from an egg. Insect larvae do not keep their outward appearance because they metamorphose into some apparently different creature whose parts appear simultaneously and complete.

In this brief comparative study of the heart in accordance with the *scala naturae*, the classification which Harvey is following is largely Aristotelian in which the main division was between the *Enaema* or sanguineous animals and the *Anaema* or bloodless animals. Aristotle had omitted the zoophytes as a separate group but they were added to the classification by Edward Wotton in his *De differentiis animalium* (Paris 1522). There seems to be no apparent reason why Harvey should list oysters and mussels as zoophytes. Both are bivalve molluscs and he refers in the next paragraph to the heart of molluscs and has already mentioned them in Chapter 4. Perhaps he did not read his notes very carefully at this point, or took his remarks from different pages.

2 The first work on the anatomy of the louse, *pediculus*, was probably done by Jan Swammerdam before 1675, but in 1696 it was dissected under the microscope by Leeuwenhoek who reported having seen an incessant to and fro movement of the blood in the stomach and gut.

3 Cf. *Anat. Lect.*, p. 297: 'Frogs and snakes do not breathe in winter';

De motu locatli animalium, p. 153: 'frogs and swallows hide in winter and give forth no movements'. Sir Kenelm Digby in his *Two Treatises* alludes to the belief current at the time that swallows hibernated in the Thames.

4 *P.A.* 665b 10–15.

5 Cf. *Anat. Lect.*, pp. 257, 251.

6 *P.A.* 666b 17; *Anat. Lect.*, p. 259. These are the chordae tendineae and the papillary muscles.

7 The 1628 text is corrupt here. It reads: *Tricuspides in introitu a vena cava, et arteria venosa ianitores. . . .* The inlet valve 'in the entrance to the pulmonary vein', the left atrioventricular valve, is the mitral valve and is bicuspid, not tricuspid. Willis did not notice this anatomical error and K. J. Franklin avoided it by omitting 'tricuspid' and paraphrasing: 'ventricular valves'. But Harvey is not here referring to the two inlet valves of the heart, but to the two valves on the right side of the heart whose action he says is less efficient than those of the left side. Moreover, he continues with a detailed description of the action of the mitral valve and its two cusps. It would seem clear, therefore, that the error here arises from a misprint, *arteria venosa* for *vena arteriosa*. The nomenclature is confusing and this error is not uncommon in anatomical texts. There is no reason to suppose that printers knew their anatomy or were aware of the mistake they were making.

8 An allusion to Aristotle's false statement that the heart has three ventricles. It greatly worried anatomists of the sixteenth century who were much concerned to explain what he had meant by it.

9 *Anat. Lect.*, p. 323.

10 This treatise was never published and Harvey's notes for it are not known to exist. A number of his observations on respiration will be found in *De generatione* and in the *Anatomical Lectures*. See the Introduction to my edition, pp. liii–lvi.

11 *De motu locali animalium*, p. 103: 'all movements start from contraction'.

12 This treatise was never written up for publication but the notes for it exist; see *Harvey's* De motu locali animalium, *1627*.

13 For my discussion of Harvey's views on the primacy of the heart and the antiquity of the blood and of his compromise with Aristotle, see *WH and the Circulation of the Blood*, pp. 227–35.

INDEX

Printed in Great Britain
at the Alden Press, Oxford
and bound by
Kemp Hall Bindery, Oxford